Permaculture
in a
Nutshell

Patrick Whitefield

Permanent Publications

Published by
Permanent Publications
Hyden House Limited
The Sustainability Centre, East Meon
Hampshire GU32 1HR, England
Tel: 01730 823 311 or 0845 458 4150 (local rate UK only)
Fax: 01730 823 322
Overseas: (int. code + 44 - 1730)
Email: enquiries@permaculture.co.uk
Web: www.permaculture.co.uk

Distributed in the USA by:
Chelsea Green Publishing Company
PO Box 428, White River Junction, VT 05001
Tel: 802 295 6300 Fax: 802 295 6444
Web: www.chelseagreen.com

First published 1993
2nd edition 1997, 3rd edition 2000, reprinted 2002
© 2002 Patrick Whitefield

The right of Patrick Whitefield to be identified as author of this work
has been asserted by him in accordance with the Copyright, Designs
and Patents Act 1988.

Designed and typeset by Tim Harland.

Cover illustration by Paul Butler.
Illustrations by Glennie Kindred with additional artwork by Terry Greenwell.

Printed by Antony Rowe Ltd., Chippenham, Wiltshire.

Printed on Elemental Chlorine Free (ECF) paper.

British Library Cataloguing in Publication Data.
A catalogue record for this book is available from the British Library.

ISBN 1 85623 003 1

Contents

The Author

Patrick Whitefield, NDA, is a permaculture teacher, designer, consultant and writer.

He grew up on a smallholding in rural Somerset and qualified in agriculture at Shuttleworth College, Bedfordshire. He then acquired farming experience in Britain, the Middle East and Africa.

He has expertise in many diverse areas. These include organic gardening, practical nature conservation and country crafts such as thatching and tipi making. He was also involved in green politics for a number of years as a prominent member of the Ecology Party.

Patrick has found that his mixed experiences have led him to the logical conclusion of permaculture and are directly relevant to his present work. He is a permaculture teacher and writer who inspires respect, affection and a good measure of action wherever he imparts his considerable knowledge.

Patrick is also author of the acclaimed book, *How To Make A Forest Garden* as well as other useful booklets such as *Practical Mulching*, *Woodland in Permaculture* and *Tipi Living*.

As well as being distributed in the USA, *Permaculture in a Nutshell* has also been translated into German, Czech, Danish and Russian.

Preface
by
Jonathon Porritt

These days, it seems that just about every politician on earth is having a bit of a problem with 'the vision thing'. So few of the old ideas any longer seem capable of delivering the goods; so many of the 'new' ideas look remarkably like the old ones recycled!

And nowhere is this more true than in the area of agriculture and food. We produce it, distribute it, and retail it all in the most hopelessly unsustainable manner – and now we are about to genetically engineer it in a similar fashion.

Permaculture stands four square against that collapse into the unsustainable. It talks about food production in a different language. Its dreams are issues in a different coinage. And for its practice, it goes back to the basics of natural systems and what it is that makes them work.

So alienated are we from these natural systems that many will undoubtedly resort to ridicule or denial as their automatic defence against such a radical alternative. Which for people like me only serves to underline the importance of its contribution to the current debate.

Putting into practice the design concepts that underpin permaculture is no mean challenge. But a new era requires hard-headed and creative thinking, and you'll find plenty of that in these pages.

Jonathon Porritt
Environmental writer and campaigner

Introduction

Permaculture is an approach to sustainable living that is spreading throughout the world, from Zimbabwe to Russia and Nepal to California. This book is an introduction to permaculture primarily for people living in Britain. It explains what permaculture is, and gives examples of how it can be applied in a variety of situations, in both town and country. It also tells you where to obtain more detailed information and how to contact people already practising permaculture.

Chapter 1

What is Permaculture?

There is a great awareness these days that we are reaching the physical limits of the Earth. We cannot go on creating pollution at the present rate, or filling our ever-growing appetite for energy and materials for ever. We are so profligate with oil and other fossil fuels that we have developed a way of producing food which consumes around ten calories of energy for every calorie contained in the food.

Changing to organic methods of food production could reduce this high input by a significant amount, as both chemical fertilisers and poisons are energy intensive. But conventional organic farming still relies heavily on machinery and the transport infrastructure, so the whole process of putting food on our plates would still consume more energy than it produced. Simple peasant agriculture can reverse the situation and yield ten calories for every one expended. The energy here is almost entirely in the form of the farmers' own labour and that of their beasts, and herein lies the fear: that our only choice is between a high-energy lifestyle and one of sheer drudgery.

But there is a third choice, called permaculture.

Permaculture includes many ideas and skills that are not unique to it; some are traditional farming practices, others involve modern science and technology. What does make it unique is that it is modelled very closely on ecosystems, which are natural communities of wild plants and animals, such as forests, meadows and marshes.

Imagine a natural forest. It has a high canopy of trees, lower layers of small trees, large shrubs, small shrubs, herb and ground layers, plus plants which are mainly below ground and climbers which occupy all levels. The production of plant material is mind-boggling compared, say, to a wheat field which is only a single layer about half a metre high.

If only the forest was made up entirely of food plants, how abundant it would be! How greatly it would out-yield the wheat field!

To achieve this great production of biomass, the forest needs no inputs but Sun, rain, and the rock from which it makes its own soil. By comparison, the wheat field is a sorry state. It needs regular ploughing, cultivating, seeding, manuring, weeding and pest control. All of these take energy, human or fossil fuel. If we could create an ecosystem like the forest, but an edible one, we could do without all that oil.

That is the basic idea of permaculture: creating edible ecosystems.

How Does It Work?

What makes the forest so productive and so self-reliant is its diversity. It is not so much the number of species that is important, but the number of useful connections between them. We have all been brought up with phrases like 'the law of the jungle' and 'the survival of the fittest' ringing in our ears, and to think of competition as the natural way that wild species interact. In fact, co-operation is just as important, especially when you look at the links between different species.

Different plants specialise in extracting different minerals from the soil and, when their leaves fall or the whole plant dies, these minerals become available to neighbouring plants. This does not happen directly, but through the work of fungi and bacteria which convert dead organic material into a form which can be absorbed by roots. Meanwhile the green plants provide the fungi and bacteria with their energy needs. Insects feed off flowers and in return pollinate the flowering plants. Many plants, such as the aromatic herbs, give off chemicals which are good for the health of their neighbours. The web of useful connections grows richer and richer as you look at it.

Some of the edible ecosystems of permaculture may actually look like a forest, for example a forest garden, in which fruit trees and bushes, herbs and vegetables are all grown together, one on top of the other. But in others the copy is not so direct, for example attaching a productive conservatory to the south side of a house. The conservatory helps to heat the house during the day and the house keeps the conservatory warm during the night, so tender food plants can be grown in winter. The building does not look like an ecosystem, but the design is based on the principle of making useful connections. This is what makes ecosystems work and it is also what makes permaculture systems work.

This can only be achieved by means of careful design. Useful connections can only be made between things if they are put in the right place relative to each other. So permaculture is first and foremost a design system. The aim is to use the power of the human brain, applied to design, to replace human brawn or fossil fuel energy and the pollution that goes with it.

Permaculture design is very much about 'wholes'. If someone tells you their farm or garden is basically conventional but there is a bit of permaculture on part of it, they are mistaken. That is not permaculture. Permaculture is a process of looking at the whole, seeing what the connections are between the different parts, and assessing how those connections can be changed so that the place can work more harmoniously. This may include introducing some new elements or methods, especially on an undeveloped site. But these changes are incidental to the process of looking at the landscape as a whole.

Although permaculture started out as **perma**nent agri**culture**, the principles on which it is based can be applied to anything we do, and now it is thought of as **perma**nent **culture**. It has grown to include: building, town planning, water supply and purification, and even commercial and financial systems. It has been described as 'designing sustainable human habitats'.

How It All Began

Permaculture is not a new idea. In many parts of the world there are people, such as the inhabitants of Kerala in southern India and the Chagga people of Tanzania, who keep gardens that are modelled very closely on the natural forest. Trees, vines, shrubs, herbs and vegetables grow together just as they do in the forest. This structure, called 'stacking' by permaculturists, enables the gardens to be far more productive than either orchards or annual vegetable gardens can be on their own, because several crops are being grown on the same spot at the same time. They provide the people with all their food, most of their medicines and fibres, some cash crop, and all on a very small area of land.

Permaculture has learnt much from traditional systems such as these, and it also incorporates many practices which have been developed in recent years. For example, organic gardening, especially of the no-dig kind, and solar technology can both be important elements in permaculture design. It is important to recognise that permaculture has no copyright on many of the ideas it holds most dear and is indebted to many co-workers in the field of creating sustainable human habitats. The specific contribution of permaculture is two-fold. Firstly, it provides the element of design, a way of putting components together for their maximum benefit. Secondly, it provides an overall framework which brings together many diverse 'green' ideas in a coherent pattern.

The word permaculture was coined by two Australians, Bill Mollison and David Holmgren, when, in 1978, they published a book called *Permaculture One*. It was an idea that had fascinated Bill for years. He had spent much of his life in the bush, both as a forestry worker and as a scientist, and the original inspiration came from the forests. He studied them, realised how they work, and said, "I could make one of these."

During the 1960s and '70s, Bill came to realise, as many of us did, that our present mainstream culture is heading down a blind alley, potentially a disastrous one. So he became involved in a lot of protesting, trying to persuade the people who are supposed to be running the world to put it to rights. After a while, he realised it was not getting him anywhere and he became convinced that real change takes place from the bottom up, not from the top down. So he gave up protesting, went home and gardened. And there permaculture was born.

Permaculture is very much about taking matters into our own hands and about making changes in our own lifestyles, rather than demanding that others do it for us. This does not mean that political action is a waste of time. There are many things which are decided at the political level and will probably continue to be so for the foreseeable future. It does mean, however, that our first reaction to any problem or challenge is not, "Something must be done!", but, "What can we do about it?"

A Sense of Ethics

At the heart of permaculture is a fundamental desire to do what we believe to be right and to be part of the solution, rather than part of the problem. In other words, a sense of ethics. The ethics of permaculture can be summed up as:

Earth care

People care

Fair shares

Earth care can be seen as enlightened self-interest: the notion that we humans must look after the Earth and all her living systems because we depend on them for survival. But on a deeper level it is the realisation that the Earth is a single living organism, and we humans are part of her, in just the same way that all the other plants and animals are. We have no more right to survive and flourish than any other species. Thus the protection of all remaining wilderness areas must be one of our highest priorities.

The human habitats created by permaculture are very much more Earth-friendly than those created by present agricultural and industrial technology. But permaculture is not about turning the whole world into a productive edible ecosystem. Far from it. By adopting permaculture, we can increase the productivity of our land to such a degree that we will need much less of it, leaving far more for wilderness.

We can help to save the dwindling remains of wilderness in the world both by campaigning and by being careful about what we consume – tropical hardwoods are an obvious example. But in Britain we no longer have any wilderness in a real sense. Every acre of the island has been profoundly affected by humans, or at least by our grazing animals. Here the richest areas for wild plants and animals are semi-natural habitats, in which humans have played an important role over many hundreds, even thousands, of years. Flower-rich meadows and coppice woodlands are examples. In these, continuing human activity, such as mowing or regularly cutting the trees, is often essential to the survival of many species of wild plants and animals.

People care is just as important as Earth care. In the past, there have been societies which were completely sustainable, but at the cost of a life of drudgery for the majority of the people. We are not talking about going back to that kind of society. We are talking about replacing both drudgery and fossil fuels with the use of intelligent design.

In fact, it is becoming increasingly clear that the technical solutions to problems are very much easier to come by than the human ones. We largely know how we need to change our agriculture and industry in order to make them sustainable. How to deal with human emotions, such as fear and greed, is less simple however, and these are what really prevent us from making progress. Permaculturists are realising more and more that we must work on people care alongside Earth care if we are to have any success in establishing sustainable human habitats. This can mean anything from teaching ourselves communication and listening skills, to designing cities which cater for real human needs.

Fair shares is a matter of acknowledging that the Earth has limits. She is not of infinite size, so our appetites cannot be infinite either. However much we recycle or buy 'environmentally friendly' products, we can never consume our way out of trouble. There is no substitute for drastically reducing our consumption of non-renewable resources. Almost everything is produced from non-renewables in our present economy, such as most of our food for a start. Renewable resources which are used faster than they can be replaced are also effectively non-renewable, for example, timber and paper at present rates of consumption.

This does not mean we should all suffer in poverty. It means that the Earth can only survive in a healthy state if we match our consumption to need, not greed. This means leaving space for other species, enough food and other resources for the poor people of the world, and a clean, well-stocked planet for future generations. In other words, taking our fair share.

If you ask most people what really makes life worth living they will say it is not material things at all, but non-material ones like love and friendship, and there is no need for a limit on these. Acknowledging the physical limits of the Earth can help to free us from the never-ending obsession for more material things, and give us more time and energy for the things that really matter.

We also need to limit our population. We in the industrialised North consume far more than the people in the poor South, in the order of 40 times as much per head by one United Nations estimate, and the amount of damage we do to the Earth is greater in proportion. So it is we in the North who most urgently need to control our population.

What causes population to grow is a very complex and controversial subject. But the changes which have taken place over the past two decades in Ladakh, in northern India, as recorded by Helena Norberg-Hodge[†], are particularly revealing. When the people of this isolated Himalayan region depended for their livelihood entirely on the local ecosystem they kept their population steady, so as to be in balance with the ecosystem. Now that the Indian government has introduced a cash economy, many Ladakhis depend on resources brought from far off and their ability to obtain them depends only on their access to money. So there is no longer any immediate need to keep population within bounds, and it is rising.

[†] *Ancient Futures – Learning from Ladakh;* Helena Norberg-Hodge; Rider; 1991.

All conventional 'development', in both North and South, is aimed at increasing people's involvement in the cash economy, replacing local production for local needs with long-distance trade. Here in the North it is called economic growth, and it increasingly separates us from the resources on which we depend for survival. Only by reconnecting ourselves with our local resources can we move towards a sustainable society.

Local Solutions to Global Problems

The Earth is enormously varied. Physical, biological and cultural conditions are never the same from one place to another. What is appropriate to one country is not necessarily appropriate to another. The principles of permaculture design are broad principles, not detailed prescriptions. They can only be used in combination with deep local knowledge, and the results will look very different from place to place.

By contrast, the conventional approach is to do away with traditional, local ways of doing things and replace them with a single, global culture. Applied to agriculture, this has been called the Green Revolution and in the short term it has greatly increased yields. But it is dependent on high fossil fuel inputs, causes pollution and is destructive of both the Earth's natural systems and human societies. It cannot be sustained.

The essence of permaculture is to work with what is already there: firstly to preserve what is best, secondly to enhance existing systems, and lastly to introduce new elements. This is a low-energy approach, making minimum changes for maximum effect, and has the least destructive impact on both natural and human communities. It applies on every scale. Not only will solutions be different from country to country, but from one locality to the next, even from one garden to the next. Subtle differences of microclimate, soil and vegetation are taken into account, and so are the differences between the needs, preferences and lifestyles of different gardeners and their families.

Chapter 2

"A Tale of Two Chickens"

The best way to look at permaculture in practice is by taking an example, and there is no better example than the permacultural way of keeping chickens. A comparison with the battery method is particularly revealing, especially when you look at the two systems in terms of how they supply the chickens' needs and how they use their outputs.

Battery Chickens

The battery chickens' food is mainly grain, grown with the use of tractors and other machinery, artificial fertilisers and poisons. All of these take a lot of energy both to produce and to use, plus a great deal of raw materials. A protein supplement is added to the grain which is often fishmeal or soya imported from poor countries where the people go short of protein. The soya beans may well be grown on land cleared from virgin forest. The feed is processed at a large centralised mill, requiring transport both from the grain farm and to the chicken farm. Water is pumped to the chicken unit via the mains. The battery house takes a lot of energy both to build and to run, including the energy needed for forced ventilation to get rid of the stale air and accumulated body heat of all the birds.

Every material need is met by the use of a great deal of energy and the creation of much pollution. The chickens' welfare needs are not met at all.

As for the outputs, only the eggs are really thought of as an output. After a productive timespan, the chickens are killed and the carcasses may go for the lowest quality meat, but the manure is considered a nuisance to be got rid of, and the idea that chickens may have other things to offer us is not even considered.

Permaculture Chickens

Permaculture chickens have much of their food grown for them where they live. The chicken run is planted with trees and shrubs which produce seed or fruit which is edible to them. No transport is involved and the food simply falls down to them. We call this a chicken forage system. Supplementary feeding may be needed at some times of the year, but a well designed system will keep this to a minimum.

The chicken forage system illustrates two of the working principles of permaculture: it is a good **relative placement** to put the food plants and the consumers in the same place; and the greatest possible use is made of **perennial plants,** such as trees and shrubs. The great thing about perennial plants is that once they are established they need little or no maintenance, unlike annuals which need a big input of work every year. In this system you do not even need to harvest the food. The chickens do it for you.

If a wheat field, an orchard or a vegetable garden are placed near the chicken run, the chickens can make useful connections with them.

If they are let into the wheat field after harvest they will eat up the ears and grains that are missed in harvesting. We humans are not going to pick them up unless we go back to the drudgery of former years, when country people were so poor they were glad of the 'gleanings'. Here the chickens are making use of a resource that would otherwise go to waste.

In the orchard chickens will help to control pests such as codling moth and fruit fly by eating the insects during that part of their life cycle they spend on the ground. This connection between chickens and orchard is useful to both parties, and that is the kind of connection we are always looking for in permaculture.

The same sort of connection can be made between the chickens and the vegetable garden. This is called a 'chicken tractor'. This is not a hundred or so chickens tied up to the front of a plough, it is using their natural inclination to peck and scratch to clear the ground of weeds and pests. They are confined on a relatively small area of ground for a short time, either by a fence or in a small, easily moved ark. In a few days they not only clear the ground and manure it, but find a part of their own food needs in the process.

The chickens' connections with the wheat field, orchard and vegetable garden illustrate two more principles of permaculture.

The first is that **every need should be met from many sources**. Just as the chickens get their food from many different sources, so should we. At present the world relies on just four plants for most of its food: rice, wheat, maize and potatoes. This makes us extremely vulnerable to crop failure if conditions should change, either due to global warming or for any other reason. We urgently need to diversify.

Chicken Ark

The second is that **every plant, animal or structure should have many functions**. Most plants and animals will yield us more than one useful output – if we have the imagination to see things like pecking and scratching as a useful output. But we can get even more variety of yields by choosing plants and animals wisely. For example, one of the shrubs we might choose as part of a chicken forage system is gorse. Not only does it produce seed that the chickens can eat, but it adds to the fertility of the soil by 'fixing' nitrogen from the air and taking it into the soil. It can also provide fuel, winter feed for cattle and horses, and flowers which bloom in every month of the year, giving food for bees and pleasure to the human eye.

Another way of putting it is that we are looking for a **multiple yield**, not just a single one as in a battery farm. In this way permaculture systems can outyield conventional ones. Even if the yield of the main product is less, the total yield is more, because we are taking many different yields at the same time.

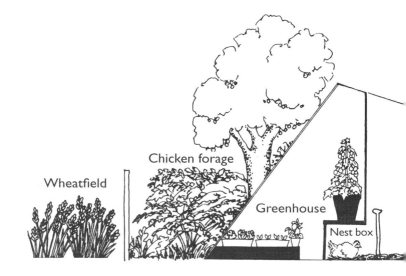

The Chicken-Greenhouse

The chicken house in a permaculture system will, as far as possible, be made out of locally produced materials. There will be a water butt to collect rainwater from the roof. This may not be enough to supply all the chickens' drinking water throughout the year, but it is a supply of water that can be had for very little outlay in energy – just a storage butt and a bit of guttering. What is more, once it is set up there is no continuing energy need for pumping. Where mains water is metered it will soon pay for itself in cash terms.

The thing that really makes the permaculture chicken house stand out is the fact that it has a greenhouse attached to the south side. The body heat of the chickens keeps the temperature up in the greenhouse at night, while the greenhouse helps to keep the chickens warm on a cold winter's morning, and the carbon dioxide breathed out by the chickens enhances plant growth in the greenhouse.

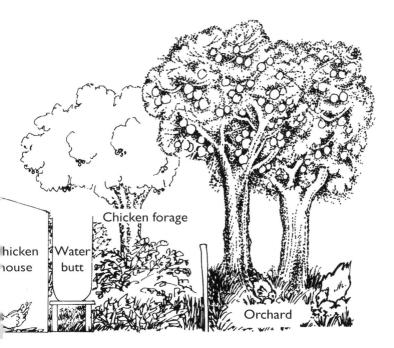

An overall pattern emerges from looking at these two ways of keeping chickens:

work = any need not met by the system

    and

    pollution = any output not used by the system

The battery system is dependent on a continuous input of energy to fulfil every need. This energy demand has been designed out of the permaculture system by making useful connections between its different parts. In the battery house the heat, carbon dioxide and manure produced by the birds are pollutants, while in the permaculture system they are useful outputs.

It is only possible to use these outputs because of the **diversity** of the system. An 'edible ecosystem', with chickens, greenhouse, vegetable garden, wheat field and orchard, can have many useful connections made between these components. A monoculture of battery hens cannot.

This kind of diversity is only possible on a **small scale**. If hundreds of thousands of birds are kept on a single farm there is no option but to feed them with bulk bought-in food, and there is no way they can be let out into the garden or the orchard to do useful work.

Using the roof to collect water is an example of another permaculture principle. Although a system may require some energy to set up, once it is running it should not need any regular energy input from outside the system. In fact, once it is established, it should **produce more energy than it consumes**. (Strictly speaking, it is impossible to produce energy. We can only change it from one form to another. But to all practical purposes the Sun's energy is unlimited and converting some of it to a usable form is a gain in real terms, whereas using up fossil fuels is a loss. Rainfall, wind and the energy in living things all come from the Sun.)

Using the body heat of the chickens, rather than paraffin or electricity to heat the greenhouse, is making use of a

biological resource. A biological resource is a plant or animal that is used to fill a need that might otherwise be filled by fossil fuels or mined minerals. The great advantage of them is that they obtain all their energy from the Sun. We, and our descendants, can go on using them for ever, whatever happens to our stocks of oil, coal and raw materials.

Harmonious Landscapes

Permaculture is very much a matter of design. Putting things in the right places is essential if any of the useful connections we are talking about are to be made. If the chicken run is next to the wheat field or orchard, the connection can be made by simply opening a gate. Only when the chicken house and greenhouse are joined together can the exchange of heat and gases take place.

These days, landscape design is usually a matter of making things look pretty rather than making them really useful. A permaculture design is primarily concerned with making the landscape productive, self-reliant and sustainable. But this does not mean that it will not be beautiful. In fact, a landscape which is designed in this way will inevitably be beautiful, just as natural ecosystems are.

Compare the sight, sound and smell of a battery house with a chicken forage system. It is like comparing a factory with an ornamental garden. In fact, there is a great deal in common between the way in which a chicken forage and an ornamental garden are laid out. Both are designed so that each plant can thrive in harmonious relationship with its neighbours, receive enough light and moisture for successful flowering and fruiting, and make the best of the particular conditions of soil and climate. What is more, many of the plants will be the same. An example is the false acacia tree, which has long been planted as an ornamental and has given its name to many an Acacia Avenue, but which also bears seed which is edible to chickens.

A permaculture landscape also tends to be an ethical one. Although it is possible to be cruel to animals in any kind of system, it is impossible not to be cruel in a battery system. On the other hand, it is very easy to be kind in a permaculture system, which is based on the idea of allowing the animals to do everything that comes naturally to them, and accepting this natural behaviour as a gift. Permaculture has no monopoly on animal rights, but they are an essential part of the system, rather than a brake that must be applied to it for pity's sake.

Chapter 3

In the City

Productive Cities

Most of us in Britain live in towns or cities. We may think there is little scope for growing food in cities, but that is just what we need to do if any kind of city life is to continue beyond the present cheap energy boom. Enormous quantities of energy are used just to transport food into the cities, and it will not be available forever. We need to grow as much of our food as we can right where we live.

The potential of parks is obvious. Over a period of time the purely ornamental trees could be replaced with kinds which are also productive, such as fruit and nut trees, giving the parks the multiple output of food as well as recreation. Allotments and city farms also show how the city can be productive.

The rest of the city-scape looks less promising until we start looking at it in three dimensions. We can train fruit trees against the walls, grow productive climbers up them and make great use of both flat roofs and balconies. This is the principle of stacking applied to an urban situation.

South-facing walls are ideal spots for most fruit trees,

especially the most tender ones such as peaches. It is even possible to grow fruit trees and shrubs in containers if the site has no soil. Morello cherries and some plums will grow against a north wall. These can be stacked with currant bushes and alpine strawberries to give high production in a shady spot. Another useful shade-tolerant perennial is the Jerusalem artichoke, and there are many others, both annual and perennial. Shady spots can also be made significantly  lighter with white-painted walls, and even mirrors, to reflect light into them.

All our common vegetables and fruit can be grown in cities, and some of the less common ones if we make use of the many sources of waste heat to be found there. What proportion of our total needs we can grow there remains to be seen, but it will certainly be far more than most of us would imagine.

People whose potential growing space is limited to a high-rise balcony, a series of window boxes or a small back yard can still grow a useful part of their food. Obviously this will not amount to much in terms of the family's bulk consumption, but it can be valuable both in terms of food value and money saving.

The difference in food value between green vegetables which are eaten within minutes of being picked, and shop-bought ones, which were picked days before, is enormous. So vegetables that are grown very close to where they are eaten have a nutritional value out of all proportion to their bulk. This is especially true of salad vegetables. A kind of lettuce which makes particularly good use of limited space is the 'Salad Bowl' type, which you pick individual leaves from rather than cutting the whole plant. If hearted lettuce is preferred, some of the miniature varieties of Cos lettuce will grow happily in a window box.

Many leafy salad vegetables can be grown on the cut-and-come-again method: seed is sown broadcast and when the seedlings come up they are clipped and allowed to sprout again. This can be repeated several times, and the repeated clippings add up to more yield than you could obtain off the same area by growing plants in rows and cutting them only once when they mature.

Seed-sprouting is a form of indoor gardening which is not to be despised. All kinds of beans, peas and lentils, seeds such as sunflower and alfalfa, and grains such as wheat, can be sprouted. As they develop from dry seeds into little plants, the food within them becomes more digestible (which effectively increases the quantity of food we can make use of), and they acquire the vitamins and essential vitality which makes fresh vegetables so much more nutritious than dried food.

Herbs have a high money value in relation to the space needed to grow them, as does garlic. So these are obvious first choices for the gardener with little space who wants a money saving from gardening. Sun loving herbs, such as thyme and rosemary, can be grown on the south side of buildings, and shade tolerant ones, like the mints and lemon balm, on the north side. The taste and health-giving properties of herbs are also much greater when they are taken fresh than when they are dried.

In some urban areas food production is limited by air pollution. Although permaculture emphasises the many things that we can do as individuals or in small groups, this is one area which can only be effectively dealt with on a political scale. In fact, lead pollution has fallen to safe levels in many areas since the introduction of unleaded petrol at a cheaper price than leaded.

Where lead pollution still is a problem, it is likely to be worst nearest the ground. So balconies and window boxes are usually okay, as are gardens or courtyards which are separated from the street by a terrace of houses. Where there is any doubt about lead, both soil and leaf analysis can be done. The local environmental health office should be able to give information on this.

Fortunately, plants have some ability to restrict their uptake of heavy metals, so we are less likely to be affected by eating city-grown vegetables than we are by breathing city air. But these pollutants still do a lot of damage, including killing soil micro-organisms and hence interfering with the soil fertility cycle.

Living Houses

The house itself is an important energy system, and in permaculture we are interested in making it more of a collector of the Sun's energy than a consumer of fossil fuels. The best way to do this is often by passive solar design. This means that the actual design of the building is such that it obtains most of its heating needs direct from the Sun, without the need for added gadgetry.

Building new houses of passive solar design is relatively easy. It costs about 5% more than conventional housing, but the extra cost is recouped in lower heating bills in around five years and after that it is pure profit for the lifetime of the building. Retrofitting an existing house is less straightforward, but there is much that can be done.

Draught-proofing, though unspectacular, gives the greatest saving in energy for the least input. Insulation comes next; a passive solar approach is to put the insulation on the outside of the walls, then the walls themselves become massive heat stores. This approach has been combined with heat recovery from the ventilation system to give a package which has reduced heating bills in high-rise flats by 90-95%.

Another good way to catch solar energy is to add a conservatory to the south side of the house. If that is not possible, it can be added to the side which catches the most sunlight, or even on the roof. Not only does this provide a highly productive growing area and extra space for the house, but the energy relationships are similar to those between the chicken house and its glasshouse. In winter, warm air from the conservatory can be vented into the main body of the house, and in summer the rising current of hot air in the conservatory can be used to draw cool air from the north side of the house and thus cool it down. Meanwhile, the conservatory converts some of the waste heat from the house into food.

Plants can be used to increase the energy efficiency of buildings. For instance, ivy grown up the north wall of a house can reduce winter heating needs. This is also beneficial to wildlife, and, if the brickwork is sound, will preserve it rather than cause deterioration. Using living plants as part of the actual structure of a building in this way is known as 'biotecture' and it is another way of using biological resources instead of non-renewables.

When using more conventional materials, it is necessary to be aware of where they come from and choose materials that respect both Earth and people. This means avoiding such materials as aluminium, with its very high energy cost, some forms of mineral fibre insulation, which may cause cancer in the workers who make them, and paints containing titanium dioxide, which is harmless in itself but very polluting in its manufacture. Alternatives to all these exist. For example, many natural insulating materials, such as wool or cork, can be preserved from rot, fire and rodents with a harmless dressing of borax.

Water supply and sewage disposal are under great strain both in cities and elsewhere, and much can be done to ease the problem by making better use of the available resources. At present, we fill the sewers with a mixture of rainwater off the roofs, 'grey' water from sinks and baths and 'black' water from toilets. The rainwater can be used for drinking, as it is probably purer than what comes out of a tap, while the grey water can be used for flushing the toilet and supplementary plant watering. By separating the three and using each quality for its highest use, mains water consumption in the city can be reduced by half for the cost of a little plumbing. In the country, rainwater could provide for all our domestic needs, as it does on many Australian farms, in a far drier climate than ours.

The real sewage can be purified by passing it through a series of carefully designed beds of reeds. Reeds and other water plants have the ability to remove organic matter, disease organisms and even chemical pollutants such as heavy metals. Reedbeds have been installed to cope with both domestic and industrial effluent. They take up less land than conventional systems, and can even be installed vertically in a series of plastic tanks.

Rather than seeing sewage as nothing but a problem we need to see it as a resource, full of valuable organic matter and plant nutrients. As long as we treat it as only something to be got rid of we will be dependent on a continuing input of fossil fuels to produce artificial fertiliser, and we will continue to pollute the seas with sewage sludge. Much work has already been done on developing safe and effective compost toilets, and these can be combined with reedbeds to make a comprehensive sewage system.

Recycling of other materials, such as paper, glass, metals and so forth, can be easier in the city than elsewhere because people are concentrated into a small area, which reduces the energy required to collect the materials. But recycling should only be our third choice, after first reducing our consumption and secondly reusing things. Returnable bottles, for example, are far more energy efficient than bottle banks and provide more employment.

This is another area where political action has a part to play. For example, in the state of Oregon in the USA, beverages can only be sold in a container which is returnable and has a deposit charged on it. We could have the same law here.

At least for the time being, a considerable proportion of the electricity supply to cities must continue to come from fossil fuels. But this can be made very much more efficient by using the heat produced by the process of generation as well as the electricity. This heat represents the majority of the energy in the original fossil fuel and it goes to waste in conventional power stations. Where small power stations are sited near where people live, the heat can be used for space and water heating. This is known as combined heat and power. It only works because of the relative placement of housing and power station.

By far the most cost-effective 'source' of electricity, however, is conservation. Reducing consumption by any means, including installing low energy light bulbs and other efficient appliances, always pays better and causes less pollution than generating more electricity.

Putting all the above ideas together would take us a long way towards creating a complete package for sustainable living. This is likely to be more than any one person or family can easily take on alone and the attitude of local councils and utility companies can range from indifferent to hostile. The answer is for local people to get together and form their own organisations for getting things done, and this is already happening in various parts of Europe, indeed all over the world.

Chapter 4

In the Garden

Garden Design

Cities certainly present the greatest challenges to permaculture designers. As we move out to the suburbs and the country the potential is greater, as most things which can be done in the cities can be done there too, and there is the added potential of bigger gardens.

Perhaps the most valuable principle used in designing a permaculture garden is that of 'zoning'. It says that things which need the most attention should be placed nearest to the centre of human activity.

How often have you seen a garden where the flowerbeds are placed near the house and the vegetables tucked away behind a hedge as far from the house as possible? This is pretty typical. Yet there is no truer saying than 'the best fertiliser is the gardener's shadow'. Vegetables grow better where you see them every day and give them the attention they need when they need it. Weeds get pulled before they start to seriously compete with the vegetables; watering gets done before the plants start to wilt.

Also, you eat more of what you grow when you can easily inspect what is ripe from day to day. It is a sad fact that every year masses of vegetables are grown and then left to rot in the garden simply because they were grown out of sight. Sometimes this is because no-one has visited the garden for a few days, so it is not known that such-and-such a vegetable is ready in quantity. At other times, you know what is there and want to cook it, but it is raining, the kids are yelling, you are behind with the cooking and the last thing you need is a trek to the bottom of the garden. A quick nip out of the back door would be quite another thing.

The most productive area of any back garden is that which can be seen from the kitchen window. The most effective way to boost the productivity of any garden is to move the vegetables into this area. You can get more food for the same work, or less work for the same food, just as you choose. Zoning can be that simple and that effective.

This need not mean losing the decorative value of the garden. Many vegetables are ornamental, such as runner beans, ruby chard, 'Salad Bowl' lettuce, ornamental kales, and most of the herbs; and many flowers, such as nasturtiums and marigolds, are edible. Also, flowers and vegetables benefit from being grown together. Not only is there a general advantage in creating maximum diversity, but there are some specific connections that are especially useful. For example, some kinds of marigolds *(Tagetes)* help to control the eelworms that prey on tomatoes and can deter weeds such as bindweed, and ground elder with the chemicals they release into the soil.

The layout is just as important to creating a beautiful food garden as the choice of plants. This is the principle behind the French *potager* style of garden where vegetables are arranged in a design which is as pleasing to the eye as to the stomach. Indeed, a well-designed vegetable/flower bed can yield as much food as the same area down to pure vegetables and be as beautiful as a pure flower bed. By accepting more than one output from the land, we double the yield.

Along with the idea of zones goes that of 'sectors'. This is a matter of placing things in relation to influences coming from beyond the garden fence. Some of these are climatic factors, such as sunshine, winds and frost. Others are more human-oriented, such as a good view or the likes and dislikes of neighbours.

The climatic factors give rise to microclimates within the garden. These are areas which have their own distinctive conditions of temperature, moisture, wind and sunlight. The amount of light reaching different parts of the garden can be particularly important in determining which plant will grow where. Temperature can also vary considerably from one part of the garden to another. Sunlight affects this, but so does the heat storage capacity of walls and other massive structures. A south-facing wall gives a choice microclimate for tender plants.

Wind can be important both in exposed gardens and in suburban ones, where the wind builds up speed funnelling through the gap between one house and the next. Careful siting of the appropriate plants or structures is needed, and a wind-break can have a multiple function if it is composed of fruiting species. Damsons, nuts and gooseberries are all fairly wind-hardy.

Before planting trees or doing any other work of a permanent nature, it is a really good idea to spend a year getting to know a garden, finding out exactly where the light and shade fall in different seasons, where the windy and sheltered spots are, and where the frost lingers in the spring. Trees last a lifetime and more. A year's careful observation and thought followed by harmonious planting is much better than a rush to get the plants into the ground followed by a lifetime's regrets.

Starting small is an excellent rule of thumb for new gardeners. There is nothing more dispiriting or frustrating than having just a little more land in cultivation than you can really manage. The garden becomes a burden and things are never quite as well done as you would like them to be. A smaller area which gets all the attention it needs can produce more than a larger area that does not.

Just how small will depend on a number of factors, including the amount of time available for gardening, the crops to be grown and so on, but an intensive vegetable bed of three by three metres could produce a very worthwhile contribution to the larder. The rest of the potential vegetable garden can be put down to a green manure crop, such as lucerne. This will provide mulch material for the vegetables at the same time as it improves the soil, and the garden can expand into this improved soil when the initial area is already running smoothly.

Low-work Gardening

There are many edible perennial plants which can be grown in the garden. As well as fruits and nuts, there are perennial vegetables, some examples of which are listed in the table opposite. A lot of the plants we normally think of as herbs can be eaten as vegetables, especially in salads, and most of them are perennial. Lemon balm, fennel and mints can be used in this way.

Many of the perennial vegetables are native plants, such as salad burnet and sorrel. The great advantage of growing wild food plants in the garden is that they really want to grow there. They have been adapted over thousands of years of evolution to thrive under local conditions. Many of our cultivated food plants are introductions from other parts of the world, and they need a lot of support from us to survive and give a yield in an environment which is basically alien to them.

Growing native plants is a way of working with nature rather than against, of accepting her gifts rather than imposing our demands. It is co-operation rather than confrontation. It both benefits the local ecology and makes life easier for us. Native plants tend to 'grow like a weed' with very little effort on our part, and though they play host to many insects and disease organisms these rarely reach pest or disease proportions as the plants have evolved to coexist with them.

One of the great advantages of perennials is that they are ready early in the spring. They spend the winter as rootstocks or bulbs, and when spring comes they are ready to burst forth above ground with masses of leaves when annual vegetables are still only seeds in the packet, or at best seedlings in the tray. In the winter, when the perennials have mostly died down, there are annuals, such as brassicas, roots, land cress and lamb's lettuce that are yielding. So a combination of perennials and annuals is a good way to get an even supply of vegetables all year round.

Some Perennial & Self-Seeding Vegetables

Name	Perennial or Self Seeder	Size: Low/Med/ Tall	Main Use
Daubentons Kale (*Brassica oleracea*)	P	M	Greens
Nine-Star Broccoli (*Brassica oleracea*)	P	T	Curds
Sea Beet* (*Beta vulgaris* ssp. *maritima*)	P	M	Greens
Chard (*Beta vulgaris* ssp. *cicla*)	S	M	Greens
Good King Henry (*Chenopodium bonus-henricus*)	P	M	Greens
Fat Hen* (*Chenopodium album*)	S	M	Greens
Sea Kale (*Crambe maritima*)	P	M	Stems
Alexanders (*Smyrnium olustratum*)	S	T	Stems
French scorzonera (*Reichardia picroides*)	P	L	Salad, mild
Pink Purslane (*Montia sibirica*)	P&S	L	Salad, mild
Salad Burnet (*Sanguisorba minor*)	P&S	L/M	Salad, mild
Chickweed* (*Stellaria media*)	S	L	Salad, mild
Lamb's Lettuce (*Valerianella locust*)	S	L	Salad, mild
Winter Purslane (*Claytonia perfoliata*)	S	L	Salad, mild
Turkish Rocket (*Bunias orientalis*)	P	M	Salad, tasty
Chicory (*Cichorium intybus*)	P	Various	Salad, tasty
Sorrel (*Rumex* spp.)	P	L/M	Salad, tasty
Land cress (*Barbarea verna*)	S	L/M	Salad, tasty
Rocket (*Eruca sativa*)	S	L/M	Salad, tasty
Hairy bittercress* (*Cardamine hirsuta*)	S	L	Salad, tasty
Nasturtium (*Tropaeolum majus*)	S	L	Salad, tasty
Ramsons* (*Allium ursinum*)	P	L/M	Leaves
Welsh Onions (*Allium fistulosum*)	P	M	Leaves
Everlasting Onions (*Allium perutile*)	P	L	Leaves
Tree Onions (*Allium cepa proliferum*)	P	M	Bulbs
Skirret (*Sium sisarum*)	P	M	Roots

Notes:
Sizes are approximate only:
Low – less than 30cm; Medium – 30-60cm; Tall – over 60cm.

* Indicates a wild plant – seed available from wildflower specialists.
Many seed catalogues list perennial vegetables under 'Herbs'.

Details of many of the above plants are given in *How To Make A Forest Garden*.

Permaculture gardeners always prefer to grow annual vegetables on a no-dig system if possible. Growing vegetables without digging has been tried and tested over many years by the members of the Henry Doubleday Research Association in their gardens all over the British Isles. Results are similar to conventional growing, though they may be better in one year and worse in another. The amount of work involved is far less, and any soil compaction can be dealt with by simply loosening the soil with a fork without digging or turning it. Potatoes can be grown without digging by covering the seed with a layer of straw or other mulch.

Annual self-seeders, which can reproduce themselves without our help, are ideal for the no-dig garden. Common examples are spinach, chard and parsley, but native annuals which are edible obviously come under this heading. Fat hen, for example, makes a good spinach, and chickweed a good base for a salad.

It is important with no-dig gardening not to walk or stand on the soil where plants are growing so as to avoid compaction. Most no-dig gardens are laid out on a raised bed system, which has alternate beds and paths, with the topsoil of the paths removed and placed on the beds. The beds are sufficiently narrow for every part of them to be reached from the paths, so that no-one need ever step on a bed. This results in beds about 1.2 metres wide and paths of about half a metre. It is a highly productive system, but almost a third of the garden is taken up by paths.

This proportion can be improved by laying out the garden in keyhole beds (*see illustration opposite*). Little paths shaped like a keyhole branch off the main one, or radiate from a central spot.

Standing in the middle of a keyhole bed or on the main path, the gardener can reach much of the ground quite easily. This area is used for plants which need the most frequent attention, such as leafy greens which are picked over a long period. A little further away, but still within reach, are plants that need less frequent attention, and further away again are plants that need little more than planting and harvesting, like garlic and onions. The last group may be out of reach without stepping onto the garden, but the odd footprint in dry weather does no harm, and stepping stones can give all-weather access.

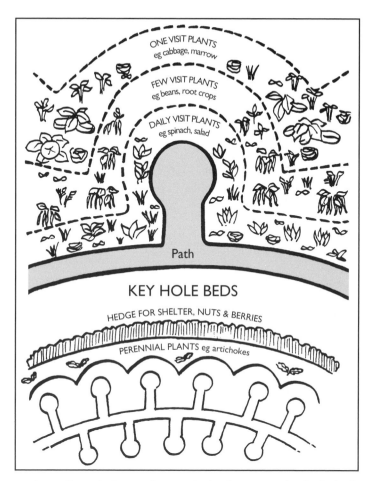

ONE VISIT PLANTS
eg cabbage, marrow

FEW VISIT PLANTS
eg beans, root crops

DAILY VISIT PLANTS
eg spinach, salad

Path

KEY HOLE BEDS

HEDGE FOR SHELTER, NUTS & BERRIES

PERENNIAL PLANTS eg artichokes

As well as their purely practical advantages, keyhole beds are more attractive to the eye than straight beds, so they lend themselves to combined edible and ornamental plantings.

Mulch is any material laid on the surface of the soil in order to kill weeds, conserve moisture and protect the soil from rain or Sun. Many mulches also add fertility to the soil as they rot down. Mulching in all its forms is a great tool of permaculture gardening. It cuts out a lot of work, and leaves the soil undisturbed, just as it would be under natural conditions.

Mulch can be used to clear new ground without digging: it kills off the existing plant cover by excluding light. Black plastic sheeting does this well, though it should never be bought new for the purpose. There is plenty of it being thrown away all the time which gardeners can recycle. Old carpets and cardboard will do as well. It takes a whole summer growing season to kill off a heavy growth of really tough plants, but if you want to grow a crop right away, and the existing growth is not too rampant, it is possible to grow plants through the mulch. *(See page 33)*

The aim of mulching is not usually to kill all the weeds but to reduce them to the sort of level that is easy to live with. In fact, the presence of a few weeds increases the diversity of the garden ecosystem and is thus very good for the health of the garden. Deep rooted ones, like dock and dandelion, bring up minerals from the subsoil. It can be worthwhile simply to chop off the leaves of these plants and use them as a nutrient-rich mulch, leaving the roots intact as a nutrient pump, rather than digging the whole plant out. Other 'weeds', as we have seen, are edible.

Weeds that send out runners, such as couch and bindweed, may not be killed by the mulch. But they tend to grow more between the mulch and the soil than in the soil itself. So at the end of the growing season you can just pull back the mulch and scoop up the majority of the roots without any digging.

In an established garden any organic material, like grass mowings, leaves or shredded paper, can be used as mulch between the plants. It can virtually eliminate weeding and save as much as 40% of watering requirements by preventing evaporation from the soil.

Slugs can be a problem with mulch - as they can without it! So it may be necessary to withhold the mulch for a while in wet weather when the plants are still young. A way of controlling slugs is to dig a pond and stock it with frogs, or to keep a few ducks which can be let into the garden now and then, both of which love to eat slugs. (Chickens should only be let into the garden for tractoring, as they make a real mess of the mulch.) Perennial plants are hardly troubled by slugs, as they do not have to pass through the vulnerable seedling stage each year.

The Forest Garden

Perhaps the most thorough-going example of a permaculture garden is the forest garden developed by Robert Hart of Shropshire. It has the layered structure of a natural forest: a canopy of fruit trees, a lower layer of dwarf fruit trees and nut bushes, a shrub layer of soft fruit, a layer of perennial herbs and vegetables at ground level, plus root vegetables and climbers.

The total production of this garden is greater than a monoculture of any one of its layers could be. This is partly because of the beneficial effect of such diversity on plant health and partly because the forest garden makes the maximum use of the resources available to it.

It makes the most of the sunlight available to it because the different layers come into leaf at different times of the year: the herb layer first, in the early spring, followed by the shrubs, and lastly the trees. Throughout the growing season there is something at the peak of its growth, making the most of the energy available from the Sun. This is something that does not happen in a single layer planting, whether of trees, shrubs or herbs. Exactly the same sequence can be observed in a natural woodland in our climate.

Maximum use is made of the whole volume of the soil because the roots of the various plants in the garden feed at different depths. In this way the stacking of different layers above ground is reproduced below ground. Some of the plants present are particularly good at accumulating certain plant nutrients from the soil, and they contribute this to the others when those parts of them which are not eaten by humans are returned to the soil. An example is comfrey, which specialises in extracting potassium from the soil.

The plants are carefully chosen to go well together. The trees of the canopy cast a relatively light shade once they come into leaf. You could not make a forest garden with a canopy of sweet chestnut, for example, which casts a heavy shade. (There is a great potential for using sweet chestnut in permaculture designs, but not in the forest garden.)

For their part, the shrubs are relatively shade tolerant, as many of them have their natural home in woodland. Hazel nuts and red and white currants can all be found wild in British woods. The varieties grown in the forest garden are cultivated ones, bred to give a higher and more reliable yield than the wild ones, but they still have enough of their ancestral qualities to stand a little shade.

Some of the shade-tolerant herbs and vegetables grown in the forest garden are also native woodland plants, for example the wild garlic or ransoms. Good King Henry, a perennial spinach, gives a good supply of greens in either shady or sunny situations, whereas parsley actually does better in light shade than in full sunlight. Some herbs have a particular part to play in controlling pests and diseases, for example the umbellifers (members of the cow parsley family) including herbs such as lovage and sweet cicely. They attract insects into the garden which prey on pests such as caterpillars and aphids.

All the plants are either perennials or self-seeders, as this is very much a no-dig garden. The soil is kept well mulched and any plants which threaten to overwhelm their neighbours, whether weeds or over-vigorous crop plants such as some mints, are either clipped back with shears or gently uprooted when the soil is soft after heavy rain. There is very little other work to do in the garden other than harvest the food.

The forest garden provides Robert Hart with fruit throughout the year, from the first gooseberry thinnings in May, to the latest keeping apples the following spring. The green food is more limited to the spring and summer seasons, so he grows a conventional garden of annual vegetables for the winter. He also has an area for sun-loving perennials, including chicories and herbs like thyme and yarrow.

There is no prescription for the ideal permaculture garden. Each must be designed according to the physical conditions on the ground and, equally important, the preferences of the individual gardener. Each of us will grow the things that we and our families like to eat and we will do it in the style we feel comfortable with. Permaculture offers many ideas that each of us can incorporate into our own style.

Making a Sheet Mulch Bed

There are many ways of using mulch in the garden. One of the most valuable ways is when starting a new garden from scratch on a site full of perennial weeds. A task that looks utterly daunting when you think of digging it becomes quite easy when you use mulch. Here is how to go about it:

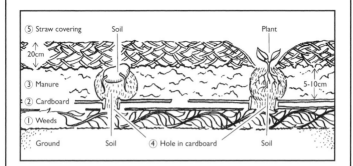

Stage 1
Knock the weeds down flat. It is not necessary to cut or remove them. A thin scattering of a high-nitrogen manure, such as blood and bone meal or chicken manure, is helpful at this stage but not essential.

Stage 2
Cover the area with a layer of cardboard, newspaper or other organic sheet material.

The purpose of this layer is to kill the weeds by excluding light. There must be no gaps and plenty of overlap between the pieces, say 20cm, to prevent vigorous weeds zig-zagging up between them. Big sheets of cardboard are best because you will get fewer joins, and an old carpet is ideal as long as it is made entirely out of natural fibres because everything you use must be able to rot down.

Newspaper is only thick enough if you use the whole thing, opened out; do not try to economise by using just a few sheets.

Stage 3
Next comes a layer to weigh the sheets down and provide some nutrients. Manure is ideal (most of our cities are ringed by riding stables which have plenty of it to get rid of), though seaweed will do, or partly rotted compost, provided it is free of weed roots and seeds. The manure does not need to be very well rotted; three months old is sufficient. This layer should be 5 to 10cm thick.

Stage 4
Now it is time to plant. Potatoes do especially well on this system. Next best are plants which are grown at wide spacings, such as marrows, sweet corn or the cabbage family. Transplants often do better than seeds, as seedlings which have only just germinated can get buried when birds come and scratch the mulch in search of grubs and insects.

Take a sharp tool, such as an old screwdriver or knife, and stab it through the mulch into the ground. This makes a hole in the sheet layer for the plant's roots to get to the soil below. Scrape away the manure from around the little hole, replace it with a couple of double handfuls of topsoil from elsewhere in the garden, and plant into this. It is not necessary to get the roots down into the soil below the sheet mulch. They will find their own way there.

Water the individual plants well, but do not water the mulch between the plants. As mulch is so efficient at conserving moisture, this is the only watering you will ever need to do, except in a very dry year.

Crops with many small seeds, like carrots, are not suitable. But remember, this system is specifically for opening up new ground. You can grow carrots on this patch next year, or specially dig a piece of ground for them if you cannot wait until then.

Stage 5

Finally, cover the bed with a layer about 20cm thick of straw or something similar. A mix of grass clippings and fallen tree leaves works well, and most local councils have plenty of both to get rid of. Hay is risky, because it may be full of seeds which will germinate and give you a big weed problem.

If you have planted potatoes you cover the whole area, but seedlings need to be left poking through. If the weather is wet, leave this layer off until the plants have grown big enough to be able to survive the attentions of slugs.

Collecting the mulch can take a little time, but it is as nothing compared to the task of digging up all those weeds and picking the pieces of root out of the soil one by one. As well as saving labour and cutting down on water use, sheet mulching is an excellent way of converting some of the detritus of the throw-away society into soil fertility.

Chapter 5

On the Farm

Do We Need Farms?

This may seem a crazy question, suggesting as it does that we could feed and clothe ourselves entirely from back gardens and smallholdings. But we could do just that if we chose.

Gardens are much more productive than farms. Research suggests that the average domestic vegetable plot in Britain yields three and a half times as much per square metre as the average farm, due to the extra attention that can be given to smaller areas. Applying the principles of permaculture design, especially those of stacking and multiple yields, can increase the productivity of the land even more.

The present strategy of large-scale mechanised farms, with only 2% of the population working on the land, is only possible thanks to the subsidy of grossly under priced fossil fuels. Imagine the last half billion years of Earth's history compressed into one year. It is now midnight on 31st December. Oil has been laid down continuously since about May. We discovered it three seconds ago and in another three seconds we will have used it all up. Whatever may be uncertain about the future, one thing is sure: we cannot go on as we are.

A permacultural vision of the future would include a far greater proportion of people growing at least some of their own food. Many of them would grow all of it and work part time for their other needs, some would grow a surplus for sale and others would grow nothing at all. There is no one model that fits everybody. But nobody will have to put in long and boring hours of labour to get their food, as permaculture replaces endless labour with skilful design.

This would enable most of the land now used for food production to be returned to its natural state, which in Britain is woodland. Some of this would supply us with timber and other produce, some would be pure wilderness, but all would provide us with two things that the world is becoming critically short of: wholesome air and clean water.

Global warming, caused by the greenhouse effect, is probably the greatest ecological threat we face, and about half of this effect is being caused by the increase of carbon dioxide in the air. We can reduce the amount of carbon dioxide we produce by burning less fossil fuels and destroying less forest; but we can also grow new forests which will convert it to solid compounds of carbon, i.e. wood.

Clean water is becoming so rare and valuable a commodity that it will soon be more expensive than petrol. Already a litre of spring water in a supermarket costs as much as a litre of petrol at the pump outside.

Water flowing from forests is clear and clean. The flow is constant, with the extremes of flood and drought evened out by the sponge effect of the vegetation. Trees also create more rain, by absorbing water and releasing it to the sky to form new clouds, rather than allowing it to flow away across the surface, taking the soil with it. By contrast, water from farmland is always contaminated to some degree and the more trees are cleared, the more droughts and floods occur.

So a permacultural vision of the future might be one where the present landscape of predominant farmland is changed to one of woodland and gardens.

But this vision is not going to happen overnight. It will only come about when a large number of us have decided to change to a more sustainable lifestyle. So for some time yet, the majority of our food will continue to come from farms, and there is a great deal we can do to make them run more efficiently and sustainably.

Farm Design

The idea of zoning, putting things which need the most attention nearest the centre of human activity, is just as important on a farm scale as it is in the garden. For example, a chicken house needs to be visited at least twice a day, while a plantation of timber trees may not need visiting even once a year, so it is clear which should be placed closest to the farmhouse. The difference between, say, a lambing shed and an orchard may be less obvious, and some careful thought about where to place them will be well rewarded. Even a very small advantage in efficiency will be multiplied up to a big gain over the lifetime of a building or of trees.

Zoning has its most obvious application when laying out a new farm or installing something new on an existing one. But it can be worthwhile to move an existing structure, if the increase in efficiency is going to be more than the cost of moving it. It can even be worthwhile expending some fossil fuel to make this possible.

An example comes from a smallholder in Wales who decided, after taking a course in permaculture, to move his polytunnel closer to the house. To do this it was necessary to create a new terrace in the hillside, and this meant using earth-moving machinery. This was a one-off expenditure of fossil fuel, an investment in a structure that will in its lifetime produce far more energy than was used to make it. It is a very different matter to the constant expenditure of energy needed to run a system that was not well designed in the first place. It is the best use we can make of our fossil fuel capital.

Sectoring, placing things in relation to influences coming from off site such as wind and sunshine, needs to be considered together with zoning. So a timber plantation, as well as being sited relatively far from the farmhouse, can also be placed to act as a windbreak. This means finding out which are the most frequent and most damaging winds on the farm before going ahead with any planting. Time spent in careful and patient observation before acting will pay for itself many times over when you are planning permanent fixtures like woods, buildings and earthworks.

Similarly, moving the polytunnel might not have been such a good idea if it meant moving it from a sunny spot to one overlooked by tall trees or a steep hill. A balance must be found between the different influences of zone and sector, and there is no rule about which will be most important on any individual site. The principles are universal, but the answers are very local.

A third factor to put in the balance is slope – planning in the vertical dimension. If there is steep land on the farm this may be the place for the timber plantation. All steep hillsides should be wooded to protect them from soil erosion. Soil erosion is not a purely tropical problem, as is often supposed. All over Britain soil is being lost from agricultural land faster than it is being replaced by natural processes.

Thought must be given to what will happen when the timber is harvested. Clear felling can leave the soil more exposed to erosion than it would have been if left under grass. But there are ways of growing timber, known as selection forestry, in which the trees are felled a few at a time, maintaining the soil cover and a healthy ecology. These are more suitable for steep slopes than is clear felling.

Slope planning is a matter of working with the landscape, rather than in spite of it. Low-lying areas prone to floods can be used for meadow, rather than arable, which is more badly affected by flooding. Buildings can be placed on gentle hillsides, preferably south-facing. Here they are above the level where cold air can accumulate overnight, causing frost pockets, and below the exposed hilltops. Water can be stored above the point where it will be used. Thus we can avoid or reduce the need for drainage works, house heating and water pumping, all of which are energy intensive.

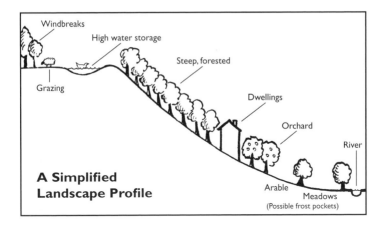

Windbreaks
High water storage
Steep, forested
Dwellings
Grazing
Orchard
River

**A Simplified
Landscape Profile**
Arable
Meadows
(Possible frost pockets)

Sector and slope planning are very much about working with microclimates. These are small areas that have their own distinctive climatic conditions, such as sunny south-facing slopes, frost pockets, sheltered areas, or windy hilltops. The first priority of good design should be to make use of existing microclimates, second comes enhancing what is already there, and only lastly should we think of creating new ones. This means matching our planting and building plans to the existing landscape, seeking out sunny spots rather than felling trees to create them, or planting willows in a marshy field rather than draining it and planting a dry land crop.

This is an energy efficient approach, as it seeks to make the minimum change for the maximum effect. It is also more in harmony with the landscape and is likely to lead to a more stable ecology.

Zone, sector and slope planning forms an integrated system for designing a farm, or any human habitat. It starts with observation of what is already there. A good designer will spend more time in listening to the landscape and its inhabitants, asking what each of them needs and has to offer, than in any other part of the design process. This observation, and the thought process that follows it, are the real work of permaculture.

Two British Farms

Permaculture systems can be designed to produce a vegan, vegetarian or meat-eating diet; the choice is a personal one. Certainly the kind of cruelty practised in factory farming is completely unacceptable. But there are some ecological advantages in having domestic animals on the farm. Apart from the obvious one of contributing to the soil fertility cycle, they can make use of resources we could not or would not use. An example already given is the chickens picking up the fallen grain from the wheat field; another is grazing animals finding a living on land that is too infertile to grow crops for our direct use. On the other hand, livestock farming is a great disadvantage when we feed them on grain because most of the grain goes to maintain the animal. We end up eating only about ten percent of the original food value in the grain.

There are already some farmers in Britain making use of permaculture principles in animal husbandry.

One of these was Bruce Marshall, who farmed sheep on the Pentland Hills in Scotland until his death in 1993. He made remarkable use of a biological resource, the earthworm. His land was typically poor hill land, boggy and acid. The conventional approach to improving such land would be to put in plastic land drains, plough, reseed, apply lime and chemical fertiliser, and continue to apply them regularly into the future. It is a high capital, high maintenance approach, and probably would not have been worth it on his land.

Instead, he chose a modest programme that involved a one-off application of rock phosphate to correct a deficiency of that mineral and lime to reduce acidity. This brought the land to a condition where clover and earthworms could survive. Then he broadcast clover seed over the existing vegetation, so it became established among the other plants and brought earthworms from more fertile soil in the valley and introduced them into the hill soil.

The clover, which can 'fix' nitrogen from the air, provides the nitrogen that would otherwise come from chemical fertiliser, while the worms keep the acidity down and improve the structure of the soil so much that artificial drainage is not needed. The beauty of the system is that it only has a modest capital cost to start up and, once running, needs no inputs from outside at all. The result was more than a doubling of the output of the farm.

Arthur Hollins is a farmer from Shropshire. He has rediscovered foggage, a traditional method of feeding cattle and sheep.

Keeping grazing animals on a conventional lowland farm can take a lot of work, as well as machinery, buildings and other inputs. Silage or hay is made in spring and summer to feed the animals through the winter. The pastures are regularly ploughed up and reseeded and artificial fertiliser applied. Cattle are usually kept inside all winter, more to protect the wet soil from the damage their feet would do to it than to protect them from the weather. This means large quantities of manure have to be shifted out onto the fields every year.

The foggage system avoids all this work. Instead of making hay or silage and hauling it to the farmyard, the winter feed is stored where it grows, in the field. Part of the farm is simply kept free of grazing for a period in the late summer to early autumn to allow the grass to grow, and the grass dries where it stands. This dried grass is the foggage and in the winter the cattle and sheep are let into this part of the farm to eat it. The only work involved is opening and shutting a gate.

The pasture is never ploughed, so over the years a strong mat of plant roots develops. This helps to protect the soil from the damaging effect of the animals' feet, so making it possible for the cattle to stay out all winter and eat the foggage where it stands.

Arthur Hollins has kept beef cattle and sheep on this system for many years and made a steady living. There is a great deal of skill involved, both in looking after the stock and in managing the grassland, but very little work. The workload is so low that a tractor is only needed to cut thistles once a year.

His farm does not produce as high an output per hectare as it would if he made hay or silage and wintered his cattle indoors, but it produces more output per unit of energy used. In fact, the amount of energy produced on this farm is many times greater than the amount consumed which is something that cannot be said about the majority of British farms.

Farming with Trees

Although Bruce Marshall has planted a lot of native trees on his improved land, neither he nor Arthur Hollins has yet taken the step of planting trees as animal fodder. Many trees have foliage which is a nutritious food for cattle and sheep, and this foliage is available in high summer, when there is often a temporary shortage of grass.

The trees can be planted out in the pasture, as in parkland or a traditional orchard, or as a hedgerow. The design will be tailor-made to fit the unique situation on each farm. The distance between the trees, or layout of the hedgerows, must be carefully chosen, so as to cause the least shading of the grass for the greatest yield of tree fodder. The choice of tree species is also important. Ash trees, for example, do not come into leaf till the grass has already done 60% of its growing for the year and they only give a light shade then. They also have leaves which are very nutritious to animals, so they are a good choice for this kind of planting.

The energy cost of transport is eliminated because the tree fodder is grown exactly where the animals live. The trees can simply be pollarded when the fodder is needed. Pollarding is cutting all the branches off a tree at the same time. The branches regrow to give the brush-headed appearance so typical of riverside willows. This ensures a regular supply of young, leafy branches.

The trees give shelter to the animals, so they can use more of their food to grow and less of it just to stay alive. Leaf fall from trees can also give a big boost to soil fertility and this can increase grass growth more than enough to compensate for the loss of production due to competition from the trees.

This combination of trees and grass makes up a simple forage system, similar to the chicken forage system. It is another example of what we mean by stacking and of taking a multiple yield. The intensive chemical farmer, using large inputs of machinery and fertilisers would produce more grass on the same area of land, but not more total produce. Skilful design has replaced the use of fossil fuel energy.

The trees need not only be for animal feed. They could also be for timber or for human food. There is great potential in Britain for growing chestnuts, walnuts and hazelnuts, quite apart from the whole range of fruits which can be grown here.

Even the acorns of our native oaks are edible, though it is necessary first to remove the tannin they contain *(See box below)*. Oaks also tend to give an irregular yield, only giving a heavy crop once in every two to seven years. But there are individual trees that yield more regularly than others and some that have 'sweet' acorns with little or no tannin in them. It would be quite possible to find these trees and selectively breed from them to produce oaks that yield sweet acorns every year.

How to Make Acorn Bread

Pick the acorns and dry them. De-husk and grind them (a coffee mill or blender will do). Put the flour in a cloth bag and pour boiling water over it to remove the tannin. Mix the resultant paste half and half with flour, and use the mix instead of pure flour in any recipe for bread, biscuits or crumble topping. It makes a rich, heavy bread, with a delicious nutty flavour.

Over the centuries, a massive plant breeding effort has gone into annual crops, especially grains. If that effort had gone into tree crops we would have trees which would far outyield the annuals. As it is, yields are comparable, but the nutritional value of nuts is much greater than that of grains, especially in protein. Already improved varieties of walnuts are available, which come into bearing much sooner and yield much more heavily than traditional varieties.

Perhaps the main limitation on nut production at the moment is that the existing varieties of chestnut and walnut will only yield well in southern parts of Britain. These species originally come from Mediterranean latitudes, but so do wheat and barley. It is only selective plant breeding which has enabled these cereals to grow in Scotland and the same can be done for the nut trees.

Combining trees and field crops, whether arable or grass, on the same land is known as agroforestry. There is experimental work going on all over Britain to develop agroforestry practices suitable for our climate and conditions. Timber and grazing, timber and cereals, and walnuts and cereals are all being tried, using conventional farming methods for the field crops. Although this kind of thing is quite far from permaculture ideals, it is a step along the way and can only be seen as a positive development.

Growing Grain Without the Plough

Tree crops, foggage farming and no-dig gardening are all examples of no-till systems. These are ways of growing food which do without ploughing or other disturbance to the soil.

No-till farming saves all the energy involved in ploughing and cultivating. What is more it preserves the natural fertility of the soil. Under natural conditions soil fertility is maintained by a host of micro-organisms, including bacteria and fungi, which recycle nutrients and make them available to plants. Most of the micro-organisms live in the top 5cm of the soil, and these die when they are buried deep by the plough. Many are sensitive to ultra-violet light and die when the soil is bared by ploughing.

Bare soil is also subject to erosion. Even when erosion is not visible to the eye it can be a constant drain on soil fertility as it is always the finest, most fertile fraction of the soil that is washed or blown away first. In its natural state soil has its own structure, a network of solid blocks or crumbs separated by fissures though which water, air and plant roots can pass. This structure is disrupted or destroyed by ploughing and replaced by an artificial structure called tilth. Tilth can be more productive than the natural structure, but it needs to be recreated every year by the plough.

So when we plough we destroy the natural fertility and structure of the soil and commit ourselves to an endless round of ploughing and manuring. The result can be a greatly increased yield per hectare, but the inputs of work, energy and materials are also greatly increased. So a no-till system usually has a higher yield per unit of energy used, in other words a higher net energy output.

This does not mean we have to change to a diet of nuts, animal products and vegetables in order to reduce our dependence on fossil fuels or hard labour. There is a no-till method of growing grains, based on the revolutionary work of a Japanese farmer called Masanobu Fukuoka.

The essence of his method is to reproduce natural conditions as closely as possible. There is no ploughing, as the seed germinates quite happily on the surface if the right conditions are provided. There is also considerable diversity, with a ground cover of clover growing under the grain plants to provide nitrogen and the weeds are also regarded as part of the ecosystem. They are periodically cut and allowed to lie on the surface, so the nutrients they contain are returned to the soil. Ducks are let into the grain plot at certain times of the year to eat slugs and other pests.

The ground is always covered. As well as the clover and weeds, there is the straw from the previous crop, used as mulch, and each grain crop is sown before the previous one is harvested. This is done by broadcasting the seed among the standing crop. Much less seed is used than in conventional growing, giving fewer, but larger and stronger plants.

In Japan, the Fukuoka method has given similar yields to chemically grown crops and much work has already been done to adapt it to European conditions. It is essentially a small-scale style of growing, suited to smallholdings, as it is one of those methods in which attention to detail replaces heavy work. It takes a great deal of skill to work with grain, clover and weeds in such a way that each fulfils its function in the system without becoming over-vigorous and crowding out one of the others. But all the work involved can easily be done by hand.

It is not suited to growing huge quantities of grain, like those presently produced in the industrialised world by means of large-scale mechanisation. But the vast majority of this grain goes to feed animals, which could be far more efficiently fed by means of diverse forage systems as outlined above. Very little is directly eaten by humans, and that amount could easily be grown by the Fukuoka method.

Nevertheless, it is unreasonable to suppose that farmers will change *en masse* to something so radically different to what they are used to in the short-term. Fortunately, there are methods now being developed which could be called a 'halfway house'. One such method is the clover/sheep/cereal system being worked on at the Institute of Grassland and Environmental Research in Britain.

This consists of growing a wheat crop in a field which also has a permanent stand of white clover. First the clover is established and, just before the grain crop is sown, it is grazed hard by sheep so that it is not sufficiently vigorous to compete seriously with the cereal. The cereal crop is sown with a machine known as a Rotaseeder, which cultivates a series of narrow strips in the turf and places the seeds in these, leaving almost all the clover undisturbed. The crop is harvested in the normal way with a slightly modified combine and sheep are let in again to prepare for the next crop.

The total amount of machine power used is much less than for a conventional crop, as one pass with a tractor replaces the three or four needed to plough, harrow and sow. Pest infestation is also much lower than on conventionally grown crops. One reason for this is believed to be the year-round cover the clover crop gives to insects which prey on the pests.

All the crop's nitrogen requirement comes from the clover. This not only saves inputs but reduces the chance of nitrate pollution in drinking water.

Nitrate pollution arises when there is more soluble nitrate in the soil than can be used by the plants growing there. A proportion of it drains to the ground water where plants can no longer reach it. The ground water forms part of our drinking supply and, although nitrates are food to plants, to humans they are poison when taken in high doses. The excess nitrate may come from artificial fertilisers, which are very soluble, or it may come from the break-up of clover-rich grassland when it is ploughed up to make way for cereals in a rotation. But growing clover and cereals simultaneously means that nitrate is never released in greater quantities than can be used by the crop. By making the right connection between clover and cereal, what was a pollutant becomes a useful input.

Again, considerable skill is required on the part of the farmer to maintain the balance between clover and grain so that they complement each other rather than compete. Weed control is more difficult than in a conventional system as many of the herbicides farmers normally use would kill the clover, and weeding by cultivation is not an option when most of the soil remains untilled. Skill is being substituted for some of the inputs of energy and chemicals.

A completely different approach to grain growing is that pioneered by Wes Jackson and his co-workers at the Land Institute in Kansas, USA. Horrified by the massive soil erosion on farmland in North America and inspired by the prairie, the natural vegetation in that part of the world, they have been working towards a 'polyculture of herbaceous perennials' for grain production.

In some ways, this is an even more radical approach than Fukuoka's. For one thing, it aims at having many different grain-producing species growing at once as in a natural prairie. For another, it turns away from the species we have traditionally grown for grain, which are all annuals, and looks for perennials which could do the job for us.

Many problems remain to be sorted out before this idea can become a viable working system, not least the problem of getting all the different plants to ripen at the same time so that they can be harvested with reasonable efficiency. But one thing is for certain: it will have the lowest inputs and the least harmful outputs of any grain growing system, and will come closer to the permacultural ideal of an edible ecosystem than any other method of growing grain.

Water on Farms

The most neglected resource on our farms is certainly water. A body of water can produce ten times the amount of protein, in the form of fish, as the same area of grazing land can in the form of sheep or cattle.

A carefully chosen selection of different kinds of fish, each making use of a different kind of food, can make full use of the diverse natural food supplies available in the pond: plant and animal plankton, vegetation, small animals such as snails, even the rich detritus at the bottom of the pond. Trees planted on the water's edge can provide much of this food in the form of falling leaves and animal life, such as caterpillars, which can rain down on the water from alder trees at certain times of the year.

Productivity can be increased by adding animal manures to the water. The most efficient way of doing this is to build pigsties or chicken houses over the pond so the droppings just fall in. This is already done in many tropical countries. It is really just the same as manuring a field, but there is less work involved and the potential return from that manure is much greater than if it was spread on the land.

The most productive ecosystems on Earth are on the edges of water. In our climate that means reedbeds, which produce more biomass than any other ecosystem and far more than any agricultural system. This edge effect can be seen in many different ecosystems, but it is most marked where water and land meet. The plants have the advantages of both mediums: the water means they never suffer from drought stress, and the soil gives them a place to root and grow that is close to the air. In order to maximise the edge effect, ponds should have a wavy shoreline, full of bays and promontories, and a shelving shore rather than a quick drop from dry land to deep water.

Many of these plants are edible and they can out-yield land plants just as fish outyield land animals. Usually the edible parts are starchy roots or tubers which are a good complement to the protein of fish. Most of them are native species, which of course makes them easier to grow and better for the local ecology. Common reed, bulrush, reedmaces and water lilies all have edible parts. As well as being harvested for human food, they can be used as the basis of pig forage systems. Pigs love the edge between water and land too as it gives them limitless opportunities to wallow!

Ponds, streams, lakes and rivers can be integrated into edible ecosystems with great advantage all round, and every farm has sites where new ponds can be created.

Water has great advantages as a source of renewable energy. Solar and wind generators only produce electricity when the Sun is shining or the wind is blowing, and this electricity must be stored in expensive batteries if it is wanted for use at another time. But water stored in a dam relatively high in the landscape can be fed to a turbine just when it is wanted, and stored as potential energy the rest of the time. On many farms there is potential to collect and store water in this way.

Where the only regular water flow is a stream passing through the lowest part of the farm there may be little scope for generating electricity with it, but this water still has great potential. There is a marvellous tool called a ram pump. With only two moving parts, it uses the energy in a stream to raise part of the flow of that stream to a storage tank much higher in the landscape. From here it can be fed by gravity to any part of the farm for a wide variety of uses: animal drinking, domestic, light industrial or garden watering.

Every farm with even a small stream running through it could probably make good use of a ram pump. It can make the mains supply redundant with energy that would otherwise flow unused to the sea.

Chapter 6

In the Community

The basic principle of permaculture is to make useful connections between different elements in a system, so that as many inputs as possible are provided from within the system, and as many of the outputs as possible are used within it. This principle can be applied to connections between human beings just as well as it can to plants and animals. In fact, sustainable human communities are possible only if they operate like this, with local needs met largely by local production.

A real community could be an urban neighbourhood or a rural village, but it needs to be small enough that people can know each other and communicate face to face. Social and emotional relationships are just as important as economic ones. Indeed, each of these affects the other and we cannot make any progress towards a more ecological lifestyle without working on all three together. Learning to communicate with each other openly and without fear is fundamental to creating true community and true communities are the essential building blocks of a sustainable world.

Today, most of us live in dormitories, whether these are country hamlets or huge housing estates. Almost all our needs are brought to us from far away and most of us travel to another place to work. We are part of an economy which is national, or global, in scale. One consequence of this is excessive energy use and pollution, as huge quantities of fossil fuels are used to shift people and things from one place to another. Another is the problem of remoteness. We become separated from the consequences of our actions, and dependent on forces way beyond our control.

When we buy something in a supermarket it is hard to know what went into producing it. The production process may involve ecological damage, ill health to the workers who make it, or cruelty to animals. By buying the product we are playing a part in that process, usually without even knowing it. When things are produced locally, by people we know, it is easy to find out how they are produced and to talk to the producer if we would like them to do it differently.

The same is true of work in the national, or global, economy. People can be thrown out of work by a decision taken on the other side of the world. Workers often have no way of knowing whether their work is part of a process that is harming the Earth and, if they do, there is precious little they can do about it. But if we are working for, or trading with, someone who lives locally, there is much more chance of having a say.

We can solve the problems of remoteness by developing communities which are self-reliant. This is not the same as total self-sufficiency; there will always be a need for some trade with other communities and other parts of the world. Self-reliant communities are ones where producing goods locally for local needs is the norm rather than the exception, where travel outside the community is a pleasure rather than a daily economic necessity, and where people are more than cogs in vast machines.

Developing this kind of community means putting power in the hands of local people, rather than national or multi-national organisations. This is not power over anyone else, but the power to decide how to run our own lives.

Consumer-Farm Links

Our long-term vision may be to replace our present large cities with many smaller settlements where the majority of people can have access to land for growing their own food. But in the short-term, most of the food eaten in cities will continue to come from farms at some distance. A useful connection can be  made by making a direct link between consumers and farmers, to the benefit of both.

These days many farmers are finding it hard to make a living, who do not benefit from the subsidy system of industrial agriculture. Farms are small businesses, so they have virtually no bargaining power when it comes to selling produce to huge companies like food processors or super-markets. When times get hard it is the farmers' incomes that get squeezed.

Most organic farmers also dislike selling their produce through supermarkets for ethical reasons. Supermarkets insist on excessive packaging, and demand high cosmetic standards, which can mean discarding as much as 50% of the produce as 'sub-standard'. Produce sold to them can also be transported hundreds of miles then sent back to be sold in the local branch a couple of miles away from the farm that grew it. Food which is treated like this can hardly be described as organic.

Meanwhile, there are many people, both in cities and elsewhere, who do not have the opportunity to grow their own food, but would like to be able to buy food they can trust at a price they can afford. Organic food is expensive in the supermarkets and no amount of labelling can substitute for knowing first hand where your food comes from.

Veggie-box schemes are a simple way of connecting farmers with producers. Each consumer agrees to buy a box of produce once a week at a standard price. They form themselves into delivery groups of about ten, the farmers deliver all the boxes to one member of each group, and the others collect from that person. Box schemes are usually for vegetables only, but they can include meat, dairy produce, flour and so on.

The farmers have an assured outlet, at something over the wholesale price, while the consumers have a supply of food they can trust at about the same price they would pay for non-organic in the shops. Transport is kept to a minimum and all packaging can be returned. The consumers are usually encouraged to visit the farms to see what is going on there and sometimes social events are organised.

Subscription farming requires somewhat more commitment than standard ordering. Each subscriber makes a single payment at the beginning of the year for the year's supply of produce. They can agree to pay by instalments, or pay in kind by working on the farm as either full or part payment.

An important advantage of subscription farming to the farmers is that they get finance when they need it, at planting time. This can help them to be free of the need to go to the banks, another situation in which farmers have little bargaining clout. The consumers can become more involved in the farm, often having a say in deciding what will be grown, and the distinction between producer and consumer begins to be less distinct.

In **Community Supported Agriculture** (CSA), the farm or farms are actually owned by the community, often in the form of a trust set up for the purpose. People who want to participate become members of a Farm Community, which may involve buying a share in the farm. All the characteristics of subscription farming are present, but the level of involvement by ordinary members in the running of the farm is much greater. This may include having regular monthly meetings at which each delivery group is represented.

The increase in mutual understanding between farmers and the people they grow food for is one of the most important benefits of CSA schemes. For city dwellers, having a place in the country where they can always go to enjoy nature, in an active as well as a passive way, can be important too. But most of

"This farm is now an integral part of all of our lives. It exists not chiefly to provide us with a living, nor to earn us fame by dramatically changing the world, but to recreate a bond between ourselves and the land. Each person has to create his or her own individual bond both with the land and also with the other members of the community. Initially we did not realise how dramatically the farm would be affected by these new practices. Usually farming today is thought of as an economic activity aiming for profit. Our aim is to produce the greatest variety of crops adapted to the needs of our local community. This concept is in fact much closer to a healthy way of farming, because an ecologically sound system has to be based on diversity."

Christina Groh, one of the founders of a community supported farm in Germany [as quoted in *The Politics of Industrial Agriculture;* Tracey Clunies-Ross and Nicholas Hildyard; Earthscan Publications].

all they provide a framework in which farming can be seen not primarily in terms of making a living, but in terms of providing a healthy living base for people, animals and plants.

In practice all systems for linking farmers and consumers are often loosely referred to as 'CSA'. Indeed, box schemes and subscription farming are really the first and second steps on the way to a full-scale CSA system. Individual systems may not fall neatly into one of the three types. In fact, it is important that each one is designed to fit the individual characteristics of the locality. Ideally, the farm or farms should be close to the urban area they link with, though in the case of large cities or remote farms this may not be possible.

The movement to link consumers and farmers started in Japan, where half a million people are now involved. It has spread to North America and some European countries, including Britain, where it is growing rapidly.

Green Money

Every time a pound is spent in a branch of a national, or multi-national, chain store, only some 20p of that pound stays in the local economy. This is mainly the wages of the shop workers. The other 80p goes straight out to pay for goods, transport, interest charges and profit. But if that pound is spent with a local person making goods out of local materials, the position could be reversed, with only 20p going outside the community to pay for materials, tools or fuels not available locally, and 80p staying in the local economy to be spent again.

The more money circulates locally, the more power goes to the local community and the less to the big faceless institutions. We become more able to see how what we are buying has been produced and what use is being made of the goods and services we are supplying others. It becomes possible to have a face-to-face relationship with the people we are trading with.

A Local Exchange and Trading System (LETS) is a currency which is only spendable locally. Having a LETSystem in a town or district encourages people to trade locally and gives them a framework in which to do so. It does not exist as coins and notes because that would be illegal, but each member of the local scheme has an account which can be debited or credited every time a transaction takes place.

Any kind of goods and services can be offered, from baby-sitting to legal advice, from home-made jam to housing. When the goods or services involve expenses which cannot be provided locally, the transaction can be in part LETS and part national currency. An example is a taxi service, where the driver's time could be paid for in 'green', as it is often called, but not the fuel.

One great advantage of LETS is that you can start spending  before you start earning. Indeed, it is essential that some people do this, because the system has no cash, so the very first transaction can only be paid for by someone going into 'commitment'. This expression is used instead of 'debt', because there is no stigma attached to having a minus balance in LETS. In fact, the total of all the minus balances inevitably must equal that of the plus balances at any one time. Without commitment the system cannot work.

This was one of the most important features in the original LETSystem, which was started in a former mining town in Canada. When the mining company pulled out, everyone's earning capacity went with it. People still had skills to offer, but no-one had the money to pay for them. It was a vicious circle, and the invention of LETS broke it.

Another effect of local currency is to even out earning levels. There was once a dentist, also in Canada, who accepted payment in LETS. Since dentists get far more per hour than most people, he eventually built up an enormous surplus, which he found impossible to spend with his more modestly-paid neighbours. His high rate of pay became meaningless, so he reduced it to something near the local average.

There are now hundreds of LETSystems worldwide, with a good number already established in Britain and many others on the way. The oldest system in Australia, and perhaps one of the most successful in the world is in Maleny, Queensland. Maleny is a small town and rural district with a population of about 7,000. 1,100 are members of the LETSystem and in 1996 they made approximately 260,000 transactions.

Community Initiatives

Local permaculture groups are springing up all over Britain. Working alone can sometimes seem daunting, and just as the different plants and animals in an ecosystem help and support each other, so do people in a group. There is no prescription for a local group. Every locality is unique and a successful group will be modelled around the specific needs of the locality and the skills of the group. An example is the group in Exmouth, Devon.

Exmouth Permaculture started by preparing sample designs for different kinds of sites in the area. An outline design was done for a farm just outside the town, including more detailed design for a chicken forage system. Outline designs were also prepared for a typical semi-detached house in the town and a local primary school. Subsequently, the group has started to install the chicken forage system and has been asked to take on some of the maintenance work at the school, which will involve implementing parts of the design.

This activity led up to the opening of the Exmouth Earth Bank, an environmental centre in the middle of the town. Services provided by the Earth Bank include: a LETSystem, a box scheme for organic food, a monthly community newsletter, permaculture courses and a drop-in information service. Rent for the premises is provided by a useful connection with some small businesses, including a solar plumber, a forest garden design and installation business, and a tax advisor. These contribute a small percentage of their earnings in return for a share of the shop-front. At present the Earth Bank is entirely staffed by volunteers, but they are looking for sponsorship to pay someone a wage.

At Kendal, in Cumbria, the local permaculturists are concentrating mainly on one project, getting some unused allotments back into cultivation.

The allotment area is an island of open space surrounded by council houses, which means they enable people to grow food right by their back doors. But until recently, only about a third of the area was being used, and the local Council were thinking of selling off the unused area for housing. Members of the local permaculture group took a step to prevent this by each applying for allotments on this site and preparing plans to work it communally. Leaflets were given to local residents inviting them to come to work days on the site.

This brought in more of the people living in the immediate area, who could see that what was going on was good. Soon they began to replace the members of the permaculture group as the main activists. As long as most of the allotment area is being used, its future is reasonably safe. In the short-term, enthusiasm for growing food may wax and wane, but it seems likely that the time is not far off when it becomes less of a pastime and more of a necessity. Then the people of Kendal will be glad that they saved the allotment site, which is just where they need it, in among the houses.

Villages, Towns and Cities

Permaculture is very much about designing landscapes as a whole, combining good housing, food production, wildlife, water supply, sewage treatment, energy supply and all other needs in an integrated way. So it is not surprising that it is being used to design whole human settlements in many parts of the world.

The pioneer permaculture settlement is Crystal Waters village in Queensland, Australia. This is a self-build village, with workplaces, agricultural land, woodland and wilderness integrated with the residential area. It is currently in the process of development, and the aim is to be 75% self-reliant in all goods and services within ten years.

The developers are a small group of people, mostly working as permaculture designers. As they did not have the money to buy land, they did a deal with a landowner whereby he exchanged the whole of his farm for a small number of developed house plots when they became ready. The people who did the design and planning work were also paid in plots. People who were interested in buying plots were invited to invest money in advance and this covered the working expenses. Thus people without any money were able to start developing land without falling into the power of the banks.

Much of what makes Crystal Waters a specifically perma-culture settlement is not necessarily visible to the casual observer. This is the care that has gone into the design. The whole site was surveyed with a thoroughness and sensitivity that went far beyond what is done for any other kind of development. All design work took place on the land, not in an office. Placement of the different components was only decided after careful consideration of all the relevant environmental factors, such as slope, aspect, soil, vegetation and microclimate, balanced with the needs of the people for space, contact with others, access and so on.

The result is a village which has been intentionally designed to work with nature rather than against. Houses are sited in the best spots for passive solar gain and summer cooling, so that the design of the house enhances the natural energy flows rather than working against them. Ponds, of which there are many, are sited within an overall plan which makes the most of the natural flows of water over the land and protects the soil from erosion. The careful placement of different kinds of vegetation encourages a steady growth in soil fertility over the years.

A major urban permaculture project is taking place in the Vesterbro area of Copenhagen, a large slum area in the inner city which is due for renewal. The authorities plan to spend US $100 million there over a ten year period, and many of the inhabitants have organised themselves into residents' groups and are demanding a say in how the development should be carried out.

Two local initiatives have been running successfully for some years already: a CSA scheme, with some 20 city people working regularly on their farm 20km from Vesterbro, and a backyard composting programme in the city. A demonstration area is planned, incorporating greenhouses for passive and active solar heating, a sewage system involving biogas generation and purification through reedbeds. These are ways of saving water and electricity and of promoting good health. One city block is to be redeveloped as the Permaculture Block. This will incorporate all the above features, and is expected to produce 10-15% of its food needs in glasshouses, courtyard and pond, as well as having a small wildlife area. A comprehensive permaculture design will also be implemented on the CSA farm.

Chapter 7

Some Questions Answered

Can permaculture feed the world?

One thing is for sure: present day industrial agriculture cannot feed the world for much longer.

Like almost all economic activity in our culture, it is utterly dependent on fossil fuels and these will soon be all but used up. Along with the consumption of fossil fuels goes a level of pollution which is equally unsustainable. The greenhouse effect, strongly associated with the burning of fossil fuels, is just one form of this pollution. What is more, present methods of food production, both in the industrialised northern hemisphere and in the peasant south, are destroying land at a frightening rate.

Soil erosion is an insidious drain on soil fertility in Europe. But when European-style farming is unthinkingly introduced into other parts of the world it becomes catastrophic. In the state of Iowa, USA, it has been calculated that for every bushel of wheat produced six bushels of soil are lost by erosion. A third of the original topsoil on cropland in the United States is already gone. This is not farming, it is mining the soil.

The spread of deserts has been speeded up both by global warming and inappropriate land use which includes over-grazing, cultivating unsuitable soils and cutting trees. It has been estimated that in the past 50 years, 50 million people have become unable to feed themselves due to desertification and another 400 million have become less able to support themselves. According to a UN study, a fifth of the world's population now live in areas which may become desertified over the next 20 years.

A third of the world's food is grown on irrigated land. Heavily irrigated land can suffer a gradual build-up of salt in the topsoil. This happens because the soil surface is frequently wet and the Sun evaporates the water from the soil leaving an ever stronger solution of salt. The salt impairs plant growth and eventually the soil becomes too salty to grow crops. Already a third of the world's irrigated land is affected by salinization and this is projected to rise to over half by the end of the decade.

The combined effects of soil erosion, desertification and salting mean that we may have to feed twice the world's present population on half the present arable land by the year 2020. Much of the lost land will be in food exporting countries, like the USA and Australia.

Permaculture offers practical solutions to these three problems. Soil erosion can be changed to soil creation by adopting no-till methods and by growing tree crops or other perennials on steep slopes. Desertification is being addressed by introducing people in arid areas to gardening, as a less destructive form of food production than extensive cropping or grazing. Re-establishing trees in arid areas is also a permaculture speciality and, once they are established, trees make their own rain.

Trees can also help stop the process of salt buildup by drawing water down through the soil towards their deep roots, away from the soil surface. But the real cure for the salt problem is to avoid causing it in the first place. This means choosing crops that suit the area, instead of trying to make the local climate suit the crops that we would like to grow. For example, we could accept a low yield of a dry-land variety of sorghum in perpetuity, rather than a few years of profitable irrigated cotton, leading to a sterile soil.

Permaculture is certainly sustainable. But how can it produce enough to feed us all?

Most of all by **gardening rather than farming**. A hectare of cultivated land in China produces nine times as many calories as a hectare in the USA, and the key to the difference is a matter of scale. Chinese farming is more what we would call gardening, based on very small plots with a lot of people involved and very little machinery. North American farming is based on very large farms with a maximum of machinery and chemicals and as few people as possible on the land.

The American system is very productive of money. (Though this mostly goes to the suppliers of the machinery and chemicals and the food processors rather than to the farmers.) But the Chinese system produces a lot of food. This is because the amount of human attention per square metre is far and away the most important factor influencing crop yield.

This is not to say that we need to use the very laborious methods that are often part of farm life in countries like China. Permaculture design and methods can take much of the work out of gardening, whether that work was formerly done by fossil fuels or human sweat. No-till methods, perennial plants, pig and chicken 'tractors' are examples.

Permaculture can also greatly increase yields, by:

. . . an increased emphasis on **aquaculture**, growing water plants and keeping fish, both of which outyield land plants and animals.

. . . taking a **multiple output**, where conventional agriculture takes only one – for example, taking the acorn crop from oak trees as well as the timber yield. This involves seeing pollutants, such as the carbon dioxide and body heat of the chicken, as potential yields. It also involves deliberately choosing plants and animals which have more than one yield. An example is the false acacia tree, which as well as being highly decorative can yield seeds for chicken forage, flowers for bee forage, leaves for cattle forage, timber which is durable without the need for preservatives, and increased soil fertility through nitrogen fixing.

. . . **stacking** two or more crops on the same piece of land, as in the forest garden. Another example is growing sweetcorn and pumpkins together. The tall thin plant and the low ground cover plant make use of different layers of the air space, so rather than competing they complement each other and make fuller use of the available sunshine.

Producing enough food is only half the picture. Taking no more than our **fair share** is equally important. Currently enough food is produced to feed everyone in the world, but people are hungry because it is unfairly distributed.

A great deal of food is imported from the poor southern countries to the rich north, often to pay for interest on their debts. Much of this is protein food, such as soya, which we feed to our farm animals. Meanwhile, people in the south go hungry. We could make a good start towards fairer shares by moderating our meat, milk and egg consumption to what we can grow from our own resources.

It may work in the tropics, but can it work here?

People often point out that in our climate only a small range of plants can be grown compared to warmer parts of the world, and that this is a limitation on permaculture. True, but it is equally a limitation on conventional agriculture. The number of plants which can be grown under either system is less in our climate. The contrast is really between the tropics and the temperate lands, not between agriculture and permaculture.

The work of Robert Hart, Arthur Hollins and Bruce Marshall is evidence enough that permaculture works here.

The idea that permaculture is more suited to the tropics sometimes arises from the misconception that it is nothing more than forest gardening. A forest garden is a direct copy of an ecosystem, but the way in which permaculture learns from ecosystems is not usually so literal. The main lesson is that what makes an ecosystem work is useful connections between its components. So a chicken-greenhouse or a field where wheat and clover grow together are just as much permaculture as a forest garden is. So are LETS and CSA, where the connections are between people rather than between plants and animals.

Permaculture is also very much a matter of working *with* the land rather than imposing our will on it; of carefully observing what the land has to offer and what it needs; of tuning in to the unique nature of each locality; and of working with these insights to prepare designs which meet the needs of both land and people. This can be done anywhere on Earth, and there is a glaring need for it here in Britain.

Isn't permaculture just organic growing?

No. Organics is a method of growing, while permaculture is a design system. They complement each other, each providing an essential component in an overall system. Nevertheless there is a certain difference in approach.

Organic farming is based on the rotation of crops, growing a different crop on each piece of ground every year. Permaculturists, on the other hand, prefer to grow a diversity of crops on the same piece of land at the same time. It is even more like a natural ecosystem, and allows more useful connections to be made between the different plants than does a rotational system.

A second difference is that the emphasis on no-till methods and perennial crops, which is central to permaculture, is missing from most organic farming. This is an essential element of a low energy strategy for the future.

Most important of all is the fact that permaculture is applicable to far more than growing food. We have already seen how its principles can be applied in the social and economic fields. In fact, they could be applied to any human activity with great advantage.

How can farmers afford the changeover to permaculture?

For many farms, the first step is to set up some form of CSA, perhaps starting with a box scheme. That way the farmers receive a greater share of the price of their produce than if they sell through conventional channels. In that case, the changeover is not a financial problem, but a direct gain.

As permaculture is very much a low-input system, a change to permacultural methods on the farm will always save money in the long term, but the change can often save money in the short-term too. For example, where foggage provides the winter feed, no hay or silage-making machinery is needed, so a farmer who changes to a foggage system will start saving on machinery costs immediately.

Nevertheless, there are some kinds of change which require an investment of money and effort that will only be repaid in the long-term. Planting trees, for animal fodder or for nuts, is an example. This kind of change could put a strain on the farm's finances.

The answer is not to try and do it all at once, but to spread the planting over a number of years. There will not be a large bill to pay in any one year and later plantings will be paid for by money made or saved by the earlier ones. A gradual approach has other advantages too: it enables the farmer to learn by experience, instead of making mistakes on the grand scale; and fewer trees planted at one time means that they can be given more attention in the first year or two after planting out, when they need it most. This is known as 'rolling permaculture'.

The farmer could also choose to involve the community in a medium to long-term version of subscription farming, with consumers investing money at the time the trees are planted in return for a share of the produce once they start bearing in a few years' time. It would be possible to buy and sell such shares, so that people who moved out of the area in the meantime would not lose out.

Changes which require expenditure, like tree planting and pond digging, can be combined with those that save money in the short-term, like reducing mechanisation and chemical use. Grants may also be available for many of the changes.

The grant system is a complicated maze with many overlapping schemes run by the Ministry of Agriculture, the Forestry Authority, local authorities, the EC and others, and what is available is constantly changing. It is always worth checking to see what is on offer. For example, at present there is a grant available from the EC towards the storage and marketing of hazelnuts, and one from the Countryside Commission for the restoration and management of traditional orchards.

Overall, the changeover can be flexible. Permaculture is not a rigid system which you either do a hundred percent or not at all. Although a permaculture design always treats the farm as a whole, it is not necessarily appropriate that every single permaculture idea be applied to that farm. Some fields may be more suitable for a no-till system than others, or some aspects of permaculture may appeal to the farmer more than others. The aim is to help people farm and live more harmoniously, not to impose a set of rules on them.

Can city-dwellers get the same benefits from permaculture?

Yes. CSA benefits consumers just as much as farmers. Starting a LETSystem is often easier in the city than in the country, as there is a larger pool of people from which a committed core group may be drawn. These social and financial aspects of permaculture may be the place to start for city people.

There is more opportunity to grow food in the city than first meets the eye. Even a small back yard and a few window sills can produce quantities of food that would astonish anyone who did not know the potential of permaculture design.

Allotments are available in many areas. In fact, the problem at the moment is often one of persuading the local council not to sell off unused allotment land, rather than finding a vacant plot. City farms come in all shapes and sizes, though they are usually more of a size to give urban people a taste of working with crops and animals rather than for significant food production. But some city farms are beginning to make links with larger areas of land in the country.

There is also a lot of unused land. Many people make little use of their gardens and would be only to pleased to have a neighbour do so, perhaps in return for a small share of the produce. Local authorities, utilities and many other bodies or individuals frequently have land which they are not using but do not want to get rid of for some reason. They are often very pleased if someone comes along and offers to manage it for them.

It is not always necessary to have access to land itself in order to produce food. It can sometimes be enough to have access to a single element in it. For example, many people have apple trees in their gardens which they never bother to harvest. The apples simply fall and rot. It would be very easy to run an 'apple clearance service' and collect all the apples for juicing. There is money to be made here, either as part of a living, or for funds for a local permaculture group. There is a man in Australia who became very rich doing just this with chestnuts.

Above all, permaculture provides a systematic approach to all aspects of ecological living. It is a framework within which a host of ideas can be evaluated and seen as part of a whole. As such, it has great value both for professionals such as architects and town planners, and for us as individuals in enabling us to decide how we want to live our lives, and to contribute on a political level to the way our communities are run.

Chapter 8

Getting Started

If you're doing any of the following, you've already started practising permaculture:

- ❀ enjoying the beauty of nature;
- ❀ growing some of your own food;
- ❀ walking, cycling or taking public transport instead of going by car;
- ❀ making decisions about what you buy on the basis of how it effects the Earth;
- ❀ reusing and recycling materials;
- ❀ supporting nature conservation.

The next step may be to:

- ❀ contact other people locally who are interested in permaculture;
- ❀ go and see a site where it's being practised;
- ❀ look for a book or two on the subject;
- ❀ get a permaculture designer to come and look at your house and garden or farm;
- ❀ attend a course.

Main Contacts

The two main contact points for all these things are the Permaculture Association and Permanent Publications.

The Permaculture Association (Britain)
BCM Permaculture Association
London WC1N 3XX
England
Tel/Fax: 0845 458 1805
(local rate UK only)
Email: office@permaculture.org.uk
Web: www.permaculture.org.uk

The Permaculture Association (Britain) is a registered charity and acts as a vehicle for connecting people, ideas, resources and projects in Britain and throughout the world. In addition to the services outlined opposite it organises a yearly convergence and get-together of interested people.

Permanent Publications
Hyden House Ltd
The Sustainability Centre
East Meon
Hampshire GU32 1HR
England
Tel: (01730) 823 311
or 0845 458 4150 (local rate UK only)
Fax: (01730) 823 322
Email: enquiries@permaculture.co.uk
Web: www.permaculture.co.uk

Permanent Publications publish *Permaculture Magazine – solutions for sustainable living* (see page 83 for further details) and books on permaculture for temperate climates. They also sell a wide range of permaculture related books, videos and products by mail order catalogue and on the web.

Local Contacts

Information on local permaculture groups and contact people throughout Britain and Ireland is available from The Permaculture Association. They send a list of local groups to all new members.

If there is no local group in your area you may wish to consider starting your own. Suggestions on how to go about this are available on request from the Association, as are lists of other members in any area of Britain specified by post code.

International Contacts

A complete international directory is available from The Permaculture Association.

Visiting Permaculture Sites

The Permaculture Association has a list of permaculture sites which are open to visitors by appointment. Sites include home gardens, smallholdings, farms, woodlands and urban projects. Some of them take WWOOFers.

Books

See 'Further Reading' on pages 76-77.

Courses

Permaculture courses are held regularly in various parts of Britain:

Details of all current permaculture courses are printed in each issue of *Permaculture Magazine* and can also be accessed at **www.permaculture.co.uk**. If you want to put on a course in your own area, the Permaculture Association can give you advice on how to go about it and a list of teachers. You are advised to check on the kind of service offered by each teacher by contacting previous clients, as there is some variation in what is offered.

The Introductory Course is usually a single weekend. It is intended for people who want to put permaculture into action immediately in their own lives, and for those who may want to go on to the Full Design Course later. It contains both information and practical exercises.

The Full Design Course comprises 72 hours' teaching and is the foundation course for people wishing to take up permaculture design work. It is equally valuable for those who want to implement permaculture in their own homes or on their own land, and for people in related professions who wish to add the permacultural perspective to their existing skills. It contains both informative and practical sessions, including actual design work on local sites. This course comes in a variety of formats: a two-week intensive, a series of weekends, or evening classes.

Specialist Courses of varying length are held for those who want to go into specific subjects in greater detail. Some examples of subjects include Forest Gardening, Third World Development, Living in Communities and Agenda 21.

Longer Courses. The above courses are normally organised privately. But longer courses based on permaculture are increasingly becoming available within the state education service. An example is the Sustainable Land Use course taught by the author at Ragman's Lane Farm in Gloucestershire.

Permaculture Designers

Design is absolutely fundamental to permaculture. The idea is to put in the maximum brainpower at the beginning, the design stage, so that things will work with the minimum of effort ever after.

Most people who are interested in permaculture prefer to learn about it themselves and do their own designing. The alternative is to get some expert help in the form of a permaculture designer. Mistakes in design can be difficult

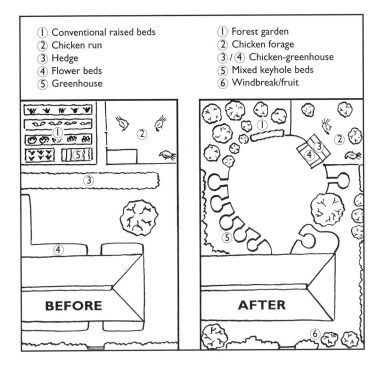

① Conventional raised beds
② Chicken run
③ Hedge
④ Flower beds
⑤ Greenhouse

① Forest garden
② Chicken forage
③/④ Chicken-greenhouse
⑤ Mixed keyhole beds
⑥ Windbreak/fruit

BEFORE

AFTER

and expensive to change at a later date. So, whether you're moving into a new place or thinking of converting your existing one to permaculture, a bit of professional help is a sound investment.

Designers work on many different levels, ranging from the "stroll round the garden and sketch on the kitchen table" to a full blown report suitable for presenting to local authorities for planning permission. Fees vary accordingly, and many designers will be open to exchanges and LETS. In addition to their training, designers have considerable experience from going round many other people's places and seeing what works and what doesn't, and this can often be their most valuable asset. Different designers have different experience and skills, some specialising in domestic-scale designs, others in farm-scale. Most have a particular subject they're especially keen on, such as woodland design or aquaculture. But whatever the kind of site, clients can expect to get back what they give in fees many times over, whether in reduced food and energy bills or increased profitability in the case of a farm. They can also expect their workload to be cut considerably.

The Association can provide a list of designers on request, and offer advice on how to choose the one who's right for you.

Further Reading

Centre For Alternative Technology Publications. A series of booklets and short books on subjects including renewable energy, energy saving in the house, water supply and sewage treatment. Practical books, for householders as much as for technologists.

Designing and Maintaining Your Edible Landscape Naturally, Robert Kourik, Metamorphic, 392pp. Brim full of practical ideas for the garden.

Earth Care Manual, Patrick Whitefield, Permanent Publications, 416pp. Comprehensive and detailed information on permaculture and it's practical application in Britain and other temperate countries.

Forest Gardening, Robert Hart, Green Books, 224pp. A wide-ranging look at forest gardening and other permacultural ideas. A good read.

Gaia's Garden, Toby Hemenway, Chelsea Green Publishing Company (USA), 260pp. A permaculture approach to gardening with much ecological information.

Hepburn Permaculture Gardens – 10 Years of Sustainable Living, David Holmgren, Holmgren Design Services, 62pp large format. An inspiring example of what one family has done to put permaculture into practice.

How to Make a Forest Garden, Patrick Whitefield, Permanent Publications, 192pp. A complete practical guide to designing and growing a forest garden.

Introduction to Permaculture, Bill Mollison & Reny Mia Slay, Tagari, 228pp. Condensed and updated version of Bill's writings on permaculture.

Local Harvest, Kate de Salincourt, Lawrence & Wishart, 256pp. An overview of the food system and a guide to different ways of linking farmers directly with consumers.

Permaculture – A Beginner's Guide, Graham Burnett, Land & Liberty, 60pp. A permaculture primer with an urban/vegan emphasis.

Permaculture – A Designers' Manual, Bill Mollison, Tagari, 576pp. The standard work, bursting with information and lavishly illustrated.

Plants For A Future – Edible and Useful Plants For a Healthier World, Ken Fern, Permanent Publications, 344 pages. Describes a very wide range of plants, mostly perennial, which are edible or otherwise useful and can be grown in temperate climates.

The Woodland Way, Ben Law, Permanent Publications, 256pp. A permaculture approach to woodland management by a practising woodsman who is committed to sustainability.

These books and many more are available by mail order from the Permanent Publications 'Earth Repair Catalogue' and online at www.permaculture.co.uk.

Useful Organisations

Association For Environment-Conscious Building, PO Box 32, Llandysul, Dyfed SA44 5ZA. Tel: 01559 370 908   Email: admin@aecb.net Web: www.aecb.net. Network of ecological and energy efficient products and contacts.

Centre for Alternative Technology, Machynlleth, Powys SY20 9AX. Tel: 01654 702 400   Email: info@cat.org.uk Web: www.cat.org.uk. Good demonstration site, including gardening, energy, water and sewage systems and much more.

Ecology Building Society, 18 Station Road, Cross Hills, Nr. Keighley, West Yorkshire BD20 7EH. Tel: 01535 635 933   Email: info@ecology.co.uk Web: www.ecology.co.uk. Lends money only on properties which are ecologically sound, often ones which mainstream building societies wouldn't touch. A good ethical place to invest any spare cash.

Environmental Transport Association, 68 High Street, Weybridge KT13 8RS. Tel: 01932 828 882   Email: eta@eta.co.uk Web: www.eta.co.uk. A road rescue service like the AA which campaigns *for* more ecological transport, rather than against.

Henry Doubleday Research Association, Ryton Organic Gardens, Woolston Lane, Ryton-on-Dunsmore, Coventry CV8 3LG. Tel: 024 7630 3517 Email: enquiry@hdra.org.uk Web: www. hdra.org.uk. Newsletter with practical information based on research and experience in organic gardening. Members can take part in nationwide experiments in their own gardens. Demonstration site.

Triodos Bank, 11 The Promenade, Clifton, Bristol BS8 3NN. Tel: 0117 973 9339   Email: enquiries@triodos.co.uk Web: www.triados.co.uk. Lends money to ethical and ecological projects. Can help with innovative financing ideas.

National Energy Foundation, National Energy Centre, Davy Avenue, Knowlhill, Milton Keynes MK5 8NG. Tel: 01908 665 555   Email: enquiries@natenergy.org.uk Web: www.natenergy.org.uk. First stop for domestic energy advice.

Federation of City Farms and Community Gardens, The Green House, Hereford Street, Bedminster, Bristol BS3 4NA. Tel: 0117 923 1800   Email: admin@farmgarden.org.uk Web: www.farmgarden.org.uk.

Plants For A Future, Blagdon Cross, Ashwater Beaworthy, Devon EX21 5DF. Tel: 01208 872 963. Does invaluable work assessing the suitability of plants for permaculture in Britain; over 1,800 species of useful plants grown on site, 12,000 on database.

Reforesting Scotland, 62-66 Newhaven Road, Edinburgh EH6 5QB. Tel: 0131 554 4321 Email: info@reforestingscotland.org Web: www.reforestingscotland.org. Working to restore the ecology of Scotland.

Soil Association, Bristol House, 40-56 Victoria Street, Bristol BS1 6BY. Tel: 0117 929 0661 Email: info@soilassociation.org Web: www.soilassociation.org. Promotes organic farming. The main national contact on community supported agriculture: keeps lists of producers looking for consumers and vice versa.

WWOOF (Willing Workers on Organic Farms), PO Box 2675, Lewes, East Sussex BN7 1RB. Tel: 01273 476 286 Email: hello@wwoof.org Web: www.wwoof.org. Puts willing workers in touch with organic farmers and gardeners in need of help, both nationally and internationally. A good way to get experience.

Suppliers

Most of the plants and seeds useful for permaculture are in common use and can be bought from fruit nurseries and seed suppliers. But some unusual plants, such as perennial vegetables or nitrogen fixing trees, can be less easy to locate. Most of them are available from at least one of the following suppliers.

Agroforestry Research Trust, 46 Hunters Moon, Dartington, Totnes, Devon TQ9 6JT. Tel: 01803 840 776 Email: mail@ agroforestry.co.uk. Seeds and plants suitable for agroforestry. Will search for unusual plants or varieties on request.

Butterworths' Organic Nurseries, Garden Cottage, Auchinleck Estate, Cumnock, Ayrshire KA182LR. 01290 551088. Apples raised to Soil Association standards.

Clive Simms, Woodhurst, Essendine, Stamford, Lincolnshire PE9 4LQ. Unusual nut trees and uncommon fruits. Mail order only. 2 x 1st class stamps for descriptive catalogue.

Cool Temperate, 5 Colville Villas, Nottingham NG1 4HN. Tel: 0115 847 8302 Email: philcorbett53@hotmail.com Web: www.cooltemperate.co.uk. Fruit, nitrogen fixers and other plants for permaculture.

Future Foods, Luckleigh Cottage, Hockworthy, Wellington, Somerset TA21 0NN. Tel/Fax: 01398 361 347 Email: enquiries @futurefoods.com Web: www.futurefoods.com. Seeds and tubers of unusual and useful plants for edible landscaping and permaculture. Also spawn of a number of edible mushroom species. For informative catalogue send 3 x 1st Class stamps.

John Chambers, 15 Westleigh Road, Barton Seagrave, Kettering NN15 5AJ. Specialist in seeds of wild plants, including edibles, dye plants etc.

Nutwood Nurseries, 2 Millbrook Cottages, Lowerton, Helston, Cornwall PL22 0NG. Nut tree specialists, with the best range in Britain. Send A5 SAE for catalogue.

Plants For A Future, Blagdon Cross, Ashwater, Beaworthy, Devon EX21 5DF. Tel: 01409 211 694. Plants of unusual perennial species for food, medicine, fuel, fibre etc.

The Willow Bank, Suite 6, Maesbury House, 7 Great Oak Street, Llanidloes SY18 6BU. Tel: 01594 861 782. Willows for basketry, shelterbelts, marginal land, soil improvement, energy crops, landscape etc. SAE for informative catalogue.

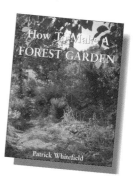

Inspired, and wondering what to do next?

Invest in yourself, and take a permaculture design course.

"The Permaculture Association supports people and projects through training, networking and research, using the ethics and principles of permaculture.

"Through our growing networks, we share skills and design sustainable solutions for the communities in which we live.

"We endeavour to be accessible to everyone in Britain and to play an active part in the developing culture of positive change."

The Association formed in 1984 to act as a focal point for education and research activities in Britain. Since then over 3,000 people have attended the design course, and the diversity, innovation and creativity of the resulting projects is breathtaking. We believe that investing in people, by helping them to develop new skills and confidences, is the best way to create sustainable systems.

By becoming a member of the Association you can contribute to our work and make use of a range of services:

* Listings of courses, projects and local groups
* The Academy, for those wishing to gain a diploma
* The Designers Register, a public listing of teachers and designers
* The Permaculture Projects Network

Membership is open to all, and donations of time, skills or money will always be put to good use. For more information contact us at:

BCM Permaculture Association, London, WC1N 3XX
Tel/Fax: 0845 458 1805 Email: office@permaculture.org.uk

www.permaculture.org.uk

The Permaculture Association (Britain) is registered with the Charity Commission for England and Wales, registration number 290897.